Sida, Botanical Miscellany 18
Botanical Research Institute of Texas, Inc.

Floristics in the New Millennium:

Proceedings of the Flora of the Southeast US Symposium

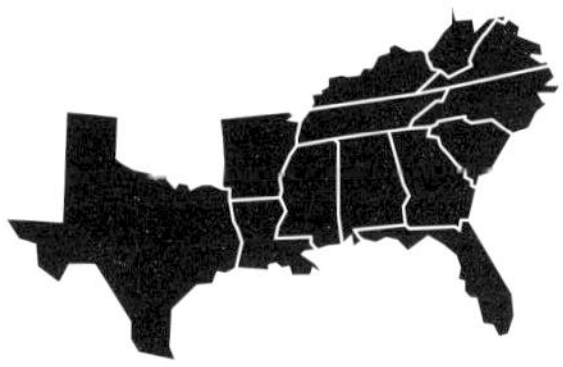

Barney L. Lipscomb
John J. Pipoly III
Roger W. Sanders
Editors

2000

ISSN 0833-1475
ISBN 1-889878-04-9

SIDA, BOTANICAL MISCELLANY NO. 18
FOUNDED BY
WM. F. MAHLER AND BARNEY L. LIPSCOMB, 1987
EDITOR: BARNEY L. LIPSCOMB
BOTANICAL RESEARCH INSTITUTE OF TEXAS, INC.
509 PECAN STREET
FORT WORTH, TEXAS 76102-4060 USA
CONTRIBUTING SPANISH EDITOR: FÉLIX LLAMAS
DPT. BIOLOGÍA VEGETAL, UNIVERSIDAD DE LEÓN
CAMPUS DE VEGAZANA S/N, 24071 LEÓN, SPAIN

COVER DESIGN/LINNY HEAGY

FLORISTICS IN THE NEW MILLENNIUM:
PROCEEDINGS OF THE FLORA OF THE SOUTHEAST US SYMPOSIUM
IS PUBLISHED WITH THE SUPPORT OF THE SID W. RICHARDSON FOUNDATION

DISTRIBUTION OF COPIES BY:

BOTANICAL RESEARCH INSTITUTE OF TEXAS
509 PECAN STREET
FORT WORTH, TEXAS 76102-4060
TELEPHONE: 1/817/332-4441
FAX: 1/817/332-4112
EMAIL: SIDA@BRIT.ORG

BRIT

PRINTED IN THE UNITED STATES OF AMERICA

Floristics in the New Millennium:

Proceedings of the Flora of the Southeast US Symposium

Table of Contents

Optimizing Underutilized Data

FOREWORD

In 1997 the Botanical Research Institute of Texas welcomed the Vanderbilt University Herbarium (VDB) to its new home at BRIT. It was a poignant move since the document that sealed this historic transaction was signed by the late Dr. Madeleine Goodman, the then Dean of the College of Arts and Science. She passed away at a much too early age of a brain tumor. It was clear to her, and her colleagues at Vanderbilt, that these wonderful collections, built primarily by Bob Kral over the years of his tenure at that University, would find an appropriate and good home in Fort Worth. To mark this historic occasion, I felt that it was appropriate to welcome the systematic botanists of the United States to a symposium celebrating the flora of the southeastern part of the country. Roger Sanders, Barney Lipscomb, John Pipoly, and other staff then made this a reality. Thus, nearly 100 botanists assembled on the 23rd, 24th, and 25th of April, 1998 for a historic gathering to focus on the flora of this part of the U.S. In so doing, we were able to announce to the world that BRIT had become the new parent, so-to-speak, of the wonderful collections that had been adopted by us, and that we took our commitment very seriously. This compilation was the result of that symposium. The proceedings serve well the memory of our evening speaker, good friend, monographer, floristician, and gentleman, Warren ("Herb") Wagner, who passed away during the intervening period.

S.H. Sohmer
Director, Botanical Research Institute of Texas
June, 2000

PREFACE

Floristics in the New Millennium: Proceedings of the Flora of the Southeast US Symposium

The Botanical Research Institute of Texas organized a Flora of the Southeast US Symposium to survey our knowledge of the flora of the SE United States, as well as to document the aspects of carrying out modern floristic work. From 23 to 25 April 1998, close to 100 botanists assembled in Fort Worth to report, discuss and summarize our current knowledge of the flora, as well as to critically examine application of new tools for conducting modern temperate floristic work. Herein we have included 14 of the papers presented at the meeting.

The Symposium was divided into three sections, summarizing background information, organizational and operational concerns in documenting the flora, and optimizing underutilized data. The first section contains review papers, including the geological history of the vegetation in the Southeast by David Dilcher, the aquatic vascular flora, including its endemism and origins, by Robert Haynes; and the region's medicinal plants by Ed Croom.

In the first section, Dilcher's paper leads off by exploring the complex history of the region's vegetation, especially through the Eocene. He proposes a near equatorial migration route for the distribution of tropical plants among North America, South America, Africa and Asia, and postulates regional extinctions to explain disjunct distributions. Following that, the paper by Haynes critically examines the definitions of aquatic plants, and chooses to define them more narrowly than has often been accepted, to include "those species of plants that grow and reproduce throughout the year in standing or flowing water." He also restricts definitions of aquatic plant habitats by eliminating any with seasonal dryness, such as flood plains of rivers, seeps, pine flat woods and wet ditches. Technical methodologies for gathering truly aquatic plants are discussed in detail. The last paper in that section is on medicinal plants by Croom, in which he traces the history of plant use in medicine throughout the region, from the early European naturalists through the present day. He makes sharp distinctions between medicinal plant use for acute vs. chronic conditions, as well as distinguishing use of autochthonous vs. introduced or naturalized species.

The next section gives us the current status of Southeast flora documentation, on a regional scale as well as to compare them with efforts across the continent. The lead paper by Funk and Morin reports on the status of herbaria housed in the Southeastern US, to discern their current statuses, activity level, and the prognosis for their long-term welfare. The results of the survey are cause for alarm, as of the 161 herbaria still active, more than 50% of the holdings are located in only seven herbaria. While 18% of the herbaria are active or very active, 47% are moderately active, leaving 35% essentially inactive. Eight herbaria, holding over 700,000 specimens, have closed and the specimens either been discarded (6,500) or transferred to other herbaria. Another 40 herbaria, totalling 25% of those surveyed, holding 1,600,0000 specimens, are either inactive or the person in charge fears that they will become so in the future. These results are, indeed, grim and the community is encouraged to place new emphasis on saving these increasingly important collections. The next paper, by Skinner and Peterson, explores the PLANTS database program, available through the USDA-Natural Resources Conservation. Among the other functions of PLANTS, the plant distributions, growth habit descriptions, vernacular names, wetland, noxious weed and endangerment status, all are features making the database extremely valuable to those engaged in local flora projects. The Morin paper addresses the Flora of North America Intranet, a rather complex network of taxon editors, regional and local specialists and others contributing to treatments for the Flora North America series. To complement the Morin paper, Barkley addresses the problems and practical editing solutions to dealing with dozens of authors when treating large families for continental floras such as the Flora North America project.

The next portion of the proceedings, "Optimizing Underutilized Data," is the largest and contains the most universally applicable information. Among the papers contained therein, almost all contain an extensive appendices, where numerous WorldWide Web sites are listed. Shulz leads off the section with a paper dealing with the many ways GIS systems can be useful in floristic studies. Following that, an exhaustive treatment of web site and unpublished data sets for the Southeast flora is dealt with by Miller and Arriagada. The next paper, by Sanders and Judd, present novel concepts for incorporation of phylogenetic data sets into floristic treatments. Then, Jackson, Grimm and Thompson summarize what is known about quaternary paleobotany and where to go for information about it. The last two papers are highly bioinformatics oriented. The first by Kartez deals with his digital flora, comprised of layered, linkable, digital databases. Finally, the volume is concluded with Robert Webster's paper on application of the Descriptive Language for Taxonomy (DELTA) systems.

In summary, we hope that the background about the state of our knowledge, status of institutional (herbarium) resources, and current activities, will encourage more floristic work, especially in the Southeastern United States. Even more importantly, we sincerely hope the last portion of this volume, with its many "how-to" and "how-to-find" papers, will permit the use of this volume as a textbook for, perhaps, a graduate seminar in floristic botany. The appendices full of web site references are meant to give this printed medium a longer, more viable lifetime.

After summing up this symposium, we would also like to add a very sincere and heartfelt welcome to the most recent addition to the BRIT Resident Research Associates, Dr. Robert Kral, and the Vanderbilt University Herbarium. The combined herbaria at BRIT, comprised of BRIT, SMU and VDB, now total some 860,000 specimens (i.e., photosynthetically-challenged plants containing LIVE data), with special strength in the central South, Southeast, and to a lesser extent, Southwestern United States. There are also significant specialty family holdings (Cyperaceae, Clusiaceae, Myrsinaceae, Xyridaceae) and strengths in focused floristic areas (Venezuelan Guayana, Colombian Chocó, the Philippines, and premontane Papua New Guinea). We feel privileged to have been host to the Flora of the Southeast US Symposium and hope you find this volume as useful as we do.

John J. Pipoly III
Roger W. Sanders
Barney L. Lipscomb
Fort Worth, Texas
July, 2000

GEOLOGICAL HISTORY OF THE VEGETATION IN SOUTHEAST UNITED STATES

DAVID DILCHER

Florida Museum of Natural History
University of Florida
Gainesville, FL 32611-7800, U.S.A.
dilcher@flmnh.ufl.edu

ABSTRACT

The present flora of southeastern United States is the product of a long history of plant migrations and intermingling of various taxa from temperate, warm temperate and subtropical origins. Floristic Provinces containing common angiosperm elements extend back about a 100 million years. These Floral Provinces display a pronounced east-west distribution that is especially well developed in the tropical, subtropical and warm temperate floras that bordered the Tethys Sea. The Tethys Sea played a major role in the distribution of tropical elements such as *Nypa* and *Hippomane*. These are found as fossils along part of their former Tethyan distribution. Regional extinctions are normal for genera that were wide ranging in the past and help explain the disjunct distributions between southeastern North America, South America and southeastern Asia. The distributions of *Podocarpus*, *Oreomunnea*, *Duguetia*, *Philodendron*, *Nyssa*, *Gordonia*, *Sabal*, *Nypa*, *Hippomane* and legumes are illustrated to demonstrate the complex history of the Eocene age vegetation in southeastern North America. A low latitude, near equatorial migration route is proposed for the distribution of tropical plants between North America, South America, Africa and Asia.

RESUMEN

La flora actual del sureste de los Estados Unidos es el producto de una larga historia de migraciones e interdigitaciones de diferentes taxa de zonas templadas, templadas cálidas y el subtrópico. Las provincias florísticas que contienen angiospermas existen desde hace 100 millones de años. Estas provincias florísticas muestran una distribución este-oeste, que está especialmente marcada en zonas tropicales, subtropicales, y templadas cálidas de los márgenes del océano de Tetis. Este océano jugó un papel importante en la distribución de algunos elementos tropicales tales como *Nypa* e *Hippomane*. Se han encontrado fósiles de ellos a todo lo largo de Tetis, aunque hoy su distribución se encuentra bastante restringida. Las extinciones regionales son bastante comunes para géneros que tenían amplios rangos en el pasado y ayudan a explicar la distribución de poblaciones actualmente aisladas, que se encuentran tanto en el sureste de Norte América como en el sureste Asiático. Las distribuciones de *Podocarpus*, *Oreomunnea*, *Duguetia*, *Philodendron*, *Nyssa*, *Gordonia*, *Sabal*, *Nypa*, *Hippomane*, y algunas Leguminosae son ilustradas para demostrar la compleja historia de la vegetación en el sureste de Norte América. Se propone una ruta de migración que ocurre en bajas latitudes para explicar parte de la distribución de plantas tropicales entre Norte América, Suramérica, África y Asia.

INTRODUCTION AND HISTORIC BACKGROUND

It is appropriate to begin this symposium with a historic view of the vegetation in southeastern North America. This history rests upon a geological foundation that consists of a history going back earlier than Texas, than the United States, or even North America (Graham 1999). I would like to give you some perspective of that history and how it relates to the past and present vegetation of the southeast United States.

The source of the data for this paper is the fossil record of plants. The questions that are asked are: what fossil plants are found in southeastern United States, where do other similar fossils occur and where do related living plants occur today? One hundred million years ago, during the Cenomanian (mid-Cretaceous), North America was divided into eastern and western landmasses, with a north-south epicontinental sea extending from the Gulf of Mexico, north through Texas, Oklahoma, Kansas, Nebraska, South Dakota and North Dakota into Canada, connecting with an epicontinental sea extending south from the Arctic Ocean (Fig. 1). This sea was 600–700 miles wide resulting in a barrier to plant migration. South America had just begun to break away from Africa. Africa had an ocean extending across portions of the Sahara Desert. Europe was a series of islands loosely connected repeatedly, through time, with Asia. India had just broken away from Africa and was migrating north through the Indian Ocean.

Each of these landmasses had Floristic Provinces as products of their history. The plant asso-

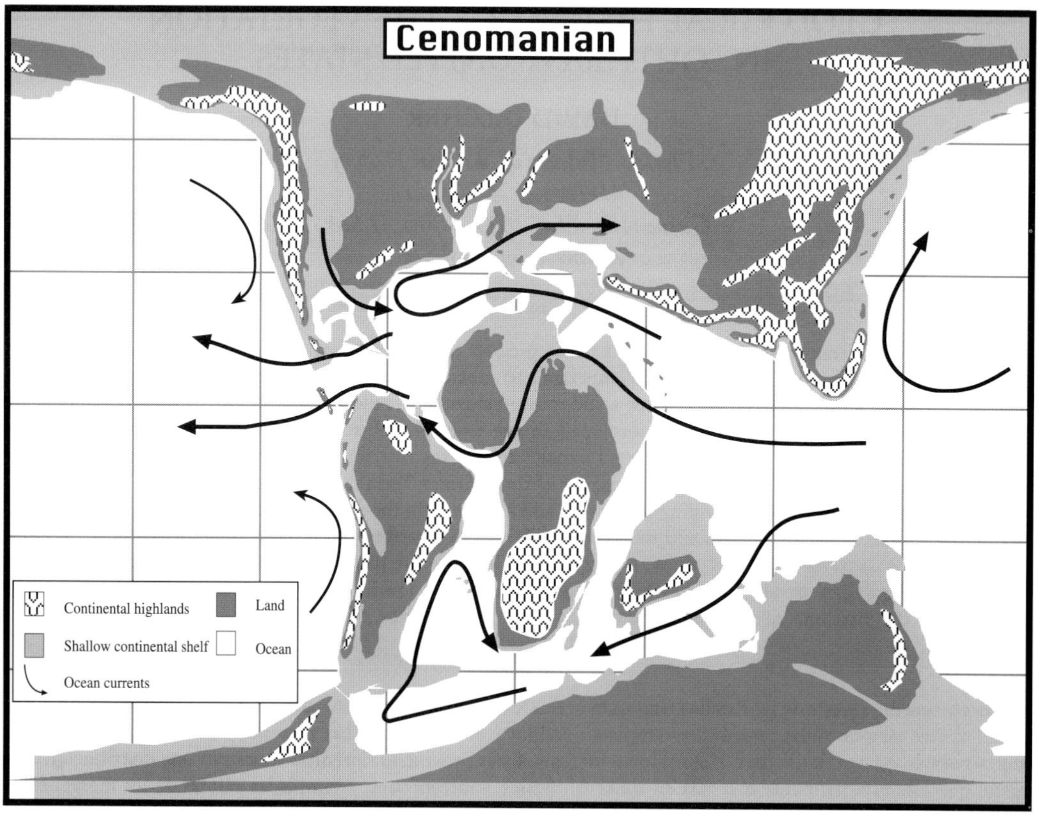

FIG. 1. Landmass map of Cenomanian showing oceans, shallow epicontinental seas, with proposed ocean currents that might carry fruits and seeds eastward along the Tethys seaway. Land and higher elevation land is also indicated. Modified from Paleomap Digital Image, Geological Society, University of Texas, 1992.

ciations that lived on them were mixtures of the plants that evolved in place and those that migrated between them. The southern most Floristic Province that we know from the Cenomanian (mid-Cretaceous), mainly from pollen data, forms a distinct Gondwana flora extending through India, South Africa, southern South America, Antarctica, Australia, New Zealand and associated islands. It is often called the Southern Gondwana Flora (Fig. 2) (Srivastava 1981, 1994). During the Upper Cretaceous across Northern Africa and Northern South America there is a common pollen flora of palms and other angiosperms forming the Elator Flora (Fig. 2). *Normapolles* type pollen typifies a common flora, shared by the connected landmass of Europe and eastern North America, while *Aquilapollenites* typifies a common flora shared through the areas of western North America and Asia. The *Aquilapollenites* and *Normapolles* Floristic Provinces continue through the Late Cretaceous, dominating the Northern Hemisphere temperate areas. This sets the stage for mixing of western and eastern floristic cold temperate and temperate elements as the epicontinental sea in North America retreats and long separated east and west landmasses merge, first in the north and finally south to the Gulf of Mexico by the early Tertiary.

Near the end of the Cretaceous, by Maastrichtian time, about 65–70 million years ago, the North Atlantic Ocean opened, India migrated north some distance and Europe was a series of islands (Fig. 3). If we look at the vegetation at this time, we see that it reflects a history stemming from the Floristic Provinces of the Cenomanian, 35 million years earlier. The Upper Cretaceous (100 to 65 million years) was a time of major angiosperm radiation. Many living families and several genera evolved during this time (Crepet 1999; Crepet & Nixon 1998; Gandolfo et al. 1998). Thus the potential for species of many extant angiosperm families to migrate around the world during the Upper Cretaceous was tremendous. During this time the low latitude east-west extending Tethys Sea opened up (Figs. 1, 3). The Tethys had a profound influence upon the

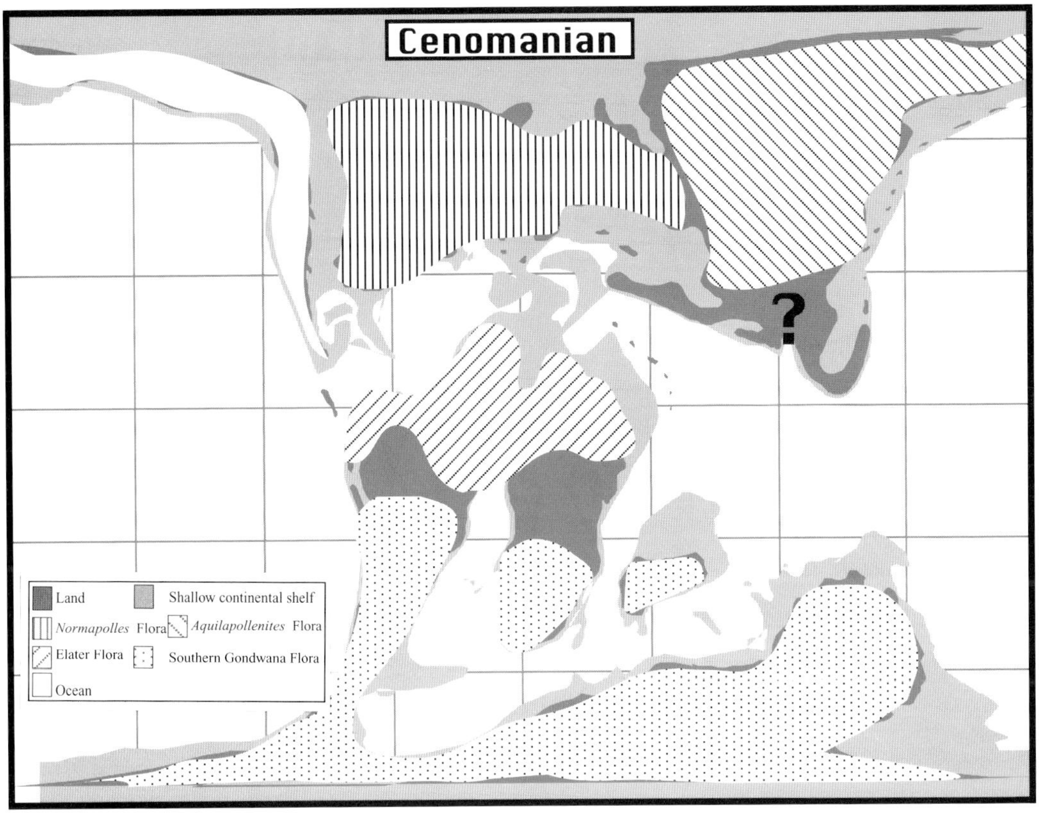

FIG. 2. Vegetational Provinces of the World during the later Cenomanian, Mid Cretaceous. Note the distinct *Normapolles* Flora in eastern North America and Europe and the Elater Flora common to northern South America and Africa. These floras border the Tethys seaway which extends east-west from Asia to the Americas. Modified from Paleomap Digital Image, Geological Society, University of Texas, 1992.

distribution of warm temperate and tropical elements of world vegetation. It influenced climates and produced ocean currents that carried plant diaspores. The "Gondwana Flora" of the Cenomanian continues as the Maastrichtian age *Podocarpus* Flora (the *Nothofagus* Flora of Srivastava 1994) in the southern part of South America, Australia, and New Zealand (Fig. 4). We don't know much about South Africa at this time (Anderson 1999) but it may have been part of this flora. India must have undergone drastic challenges to its vegetation as it moved across climatic zones, into and through the Tethys Sea. Also, there are new tropical elements that evolved along the north and south margins of the extensive east-west tropical Tethys Sea. This is a seaway that extended from Asia through the Middle East, continuing west between the islands of Europe and northern Africa and continuing on further west to the Gulf of Mexico, the Caribbean Sea and eastern North America. Tropical elements migrated along the landmasses bordering the Tethys Sea (Axelrod 1970). In the Northern Hemisphere the most northern floristic provinces are the *Aquilapollenites* Flora in Asia and western North America, and the *Normapolles* Flora in eastern North America and Europe which formed long lasting plant associations. Vegetational history of the world during the early Tertiary is founded upon all of these earlier major Floral Provinces. Tertiary vegetation of the world is a continuation of the interactions and migrations of the elements between these Maastrichtian Floral Provinces.

Pantropical Flora extended from Asia through northern Africa into northern South America and shared many tropical elements (Fig. 4) between these areas during the Upper Cretaceous and early Tertiary. At the same time a subtropical/warm temperate flora, the Laurasian Flora, extended from Asia westward through the European Islands, continuing west across what will become partially open North Atlantic Ocean on to eastern and to southeastern North America.

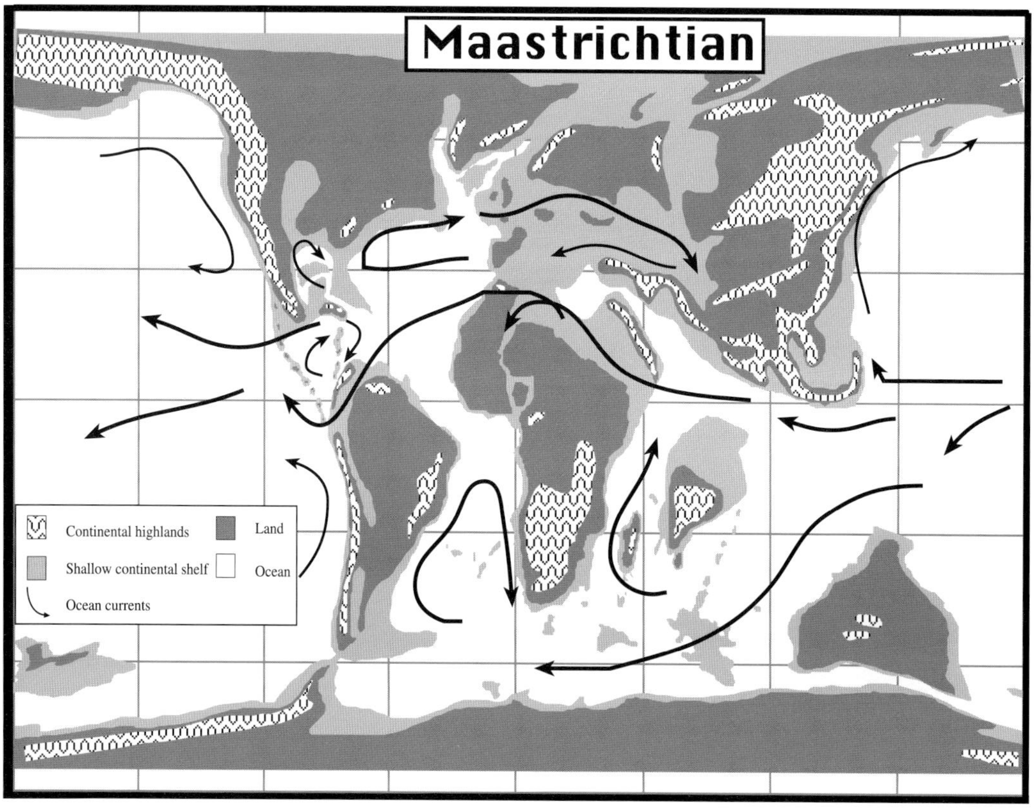

Fig. 3. Landmass map of the Upper Cretaceous, Maastrichtian showing oceans, shallow epicontinental seas and proposed ocean currents. The eastward ocean currents in the Tethys seaway would move propogules of plants lining along the lands bordering this seaway to form a common north Tethys, Laurasian Flora and south Tethys, Pantropical Flora. Modified from Paleomap Digital Image, Geological Society, University of Texas, 1992.

This Laurasian Flora mixed with elements of floras to the north and west leading to complex floral associations containing diverse floristic elements (Fig. 4). Also to the north we see a mixture of the *Normapolles* Flora with the *Aquilapollenites* Flora blurring the boundaries of these floristic provinces and forming a Pantemperate Flora of the Northern Hemisphere which became a common circumpolar boreal forest during the Tertiary (Manchester 1999). The Cretaceous Floristic Provinces presented in Figures 2 and 4 are sources for the angiosperm floras of the Tertiary. They form the foundations for the distributions of plants that we find in the Paleogene, discussed in this paper, and the distributions of many angiosperms living today.

An example of Tertiary floras in the far north, where today we have an arctic Tundra flora, is the Eocene flora from Ellesmere Island (Basinger 1986). Near 80° north latitude Basinger found fossils from the Paleocene and Eocene that produced short shoots, leaves and reproductive organs similar to the living Asian genus, *Cercidiphyllum*. This is an Asian element now but it once was abundant and widespread around the Northern Hemisphere during the early Tertiary. *Cercidiphyllum* was growing in a forest with birch and conifers such as *Metasequoia* and *Glyptostrobus*. Affinities between the Tertiary floras of Europe and living plants of Asia can be seen in such temperate genera as *Pterocarya*, *Quercus*, *Zelkova*, *Eucommia*, *Magnolia*, *Corylopsis*, *Phellodendron*, *Ailanthus* and *Actinidia* (Tralau 1963). There are numerous other records of similar fossil elements that have spread around the Northern Hemisphere, some of them by way of Alaska through Beringia into Asia, others over the north Atlantic by way of Greenland (Manchester 1999; Tiffney 1985) at latitudes about 70° N. This mixture of cool temperate and temperate plants must have dispersed by way of northern routes warmed by the Tertiary climate, warm enough to permit their growth, and allow their migration around the northern parts of the Northern Hemisphere. These routes probably were not frost free the entire year. There are many tropical and subtropi-

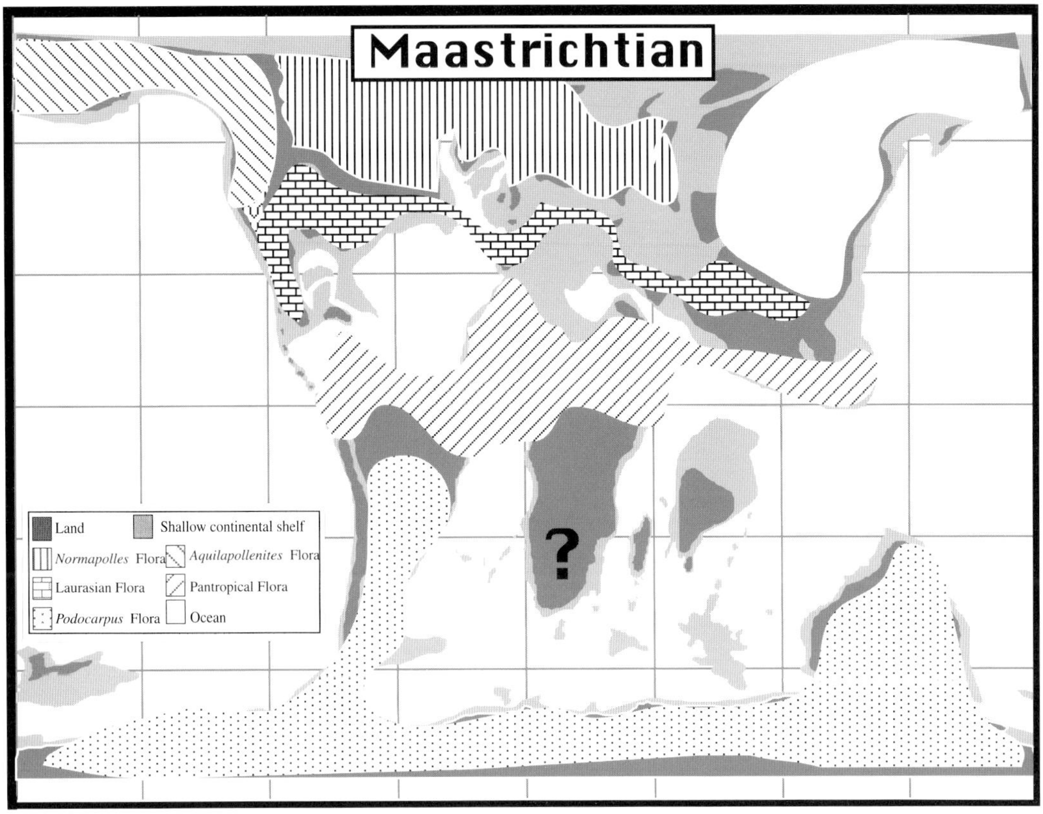

FIG. 4. Vegetational Provinces of the World during the Maastrichtian, Upper Cretaceous. This demonstrates the Laurasian Flora extending from warm temperate to tropical Asia to North America, and the Pantropical Flora also extending from Asia to South America and the Caribbean. Both of these floras border the Tethys seaway. Note the *Normapolles* and *Aquilapollenites* Flora continue to dominate the temperate and cool temperate areas of the Northern Hemisphere. Modified from Paleomap Digital Image, Geological Society, University of Texas, 1992.

cal angiosperm genera and families that are shared between Asia, Africa and Central and South America. Several of these are found as fossils in the Tertiary of southeastern United States. I will focus this paper upon a mixture of tropical, subtropical and warm temperate floristic elements from the Eocene floras in southeastern United States in order to explore the possibility that some of these plants were dispersed as part of the floras that migrated along the Tethyan Sea, linking the tropical and warm temperate areas of Asia, Africa and the Americas floristically (Fig. 24). There is evidence from the pollen and spore record that 11.5% of the species are shared between Africa and northern South America in the Paleocene. In the Eocene, new introductions of species are shared again to the extent of 11%. However, when the Paloecene floras of northern South America and the southeast United States are examined only 0.7% are shared while in the Eocene, 5.2% are shared (Jaramillo and Dilcher in press) (Fig. 24). Therefore, there is exchange and migration of floristic elements through these low tropical latitudes allowing movement between Asia, European Islands, Africa, northern South America, Central America and southeastern North America that continues into the Paleocene and Eocene.

The shoreline of southeastern North America and the Gulf of Mexico, as it existed about 45 million years ago, can clearly be seen in Figure 5. During the lower and middle Eocene there were repeated transgressions and regressions of the sea in the low-lying areas known as the Mississippi Embayment with a subtropical monsoonal climate (Dilcher 1973). During these cycles, a tremendous amount of sediment was deposited by the existing river systems, occupying similar drainage patterns as the Missouri, Ohio and Mississippi Rivers today. That is when the clays of the Claiborne Formation (Fig. 5) were deposited in western Kentucky, western Tennessee and Mississippi, in which so many plant fossils are preserved. Many of these clay

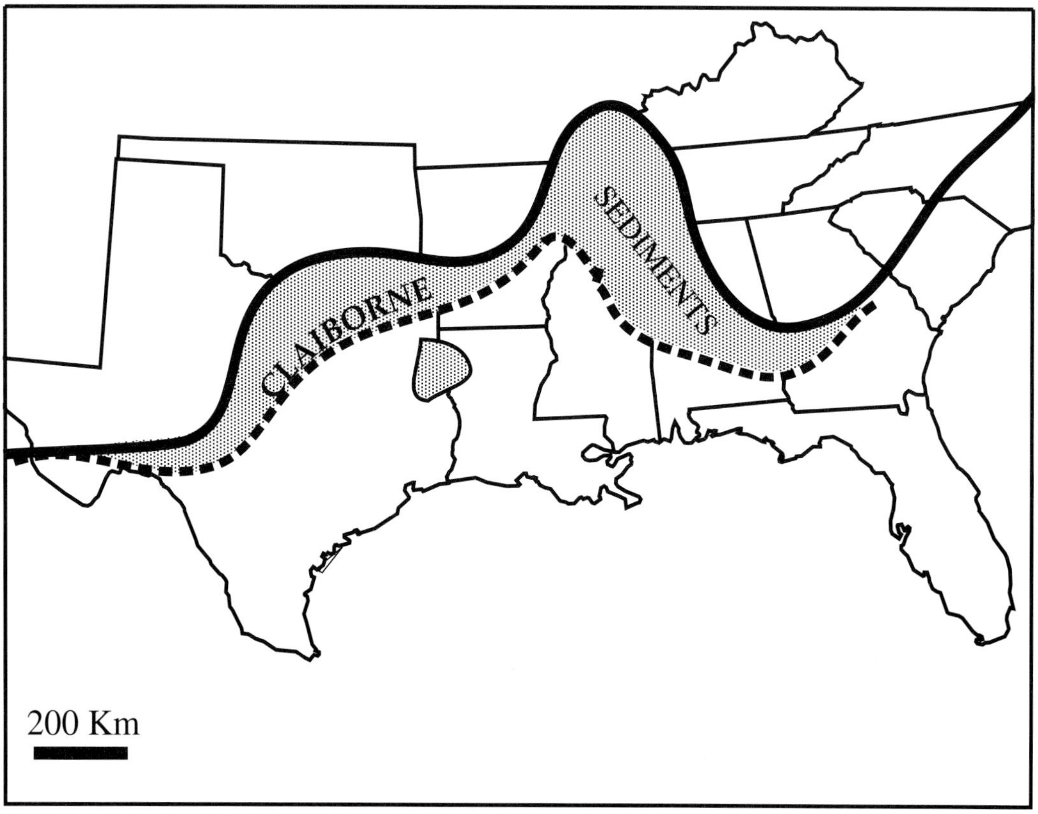

FIG. 5. Extent of middle Eocene sediments, Claiborne Formation, in southeastern United States. Modified from Murray 1961.

deposits were formed as clay plugs in ancient oxbow lakes in the flood plain of the Eocene age rivers draining this area (Dilcher 1971; Potter & Dilcher 1980).

These fossil floras collected from clay pits (Fig. 6) were first described about 80 years ago by E.W. Berry (1916, 1922, 1924, 1930, 1941). Nearly all the fossil leaves were identified to living genera. When I started examining the cuticles of these fossil leaves, I found that at least 60% of the generic identifications were incorrect (Dilcher 1963, 1974) and fossil leaves had traditionally been identified as modern genera using features only of gross form. There was no allowance for evolution by earlier researchers, which would have become obvious if the cuticular and fine venation features had been examined (Dilcher 1974). Looking at the epidermal cells of the fossil leaves, and the nature of the cuticle of extant plants of the same genus, it became apparent that many leaves were from extinct genera or species. Angiosperm paleobotanists were most often leaf matchers before about 1960 and were more concerned about paleoecology than evolution of the fossil plants they studied. They were good at matching fossil leaf forms to modern leaves and nearly always found a modern leaf to match the fossil. But they didn't look very closely at the leaf anatomy and morphology. I was trained as an anatomist and morphologist, therefore I tended to include these elements in my study of these Eocene floras and found that extensive revisions needed to be made and a new philosophical approach was needed when studying angiosperm leaf remains (Dilcher 1974).

A lot of the clay in these clay pits is what workers call rotten clay. It is so full of organic material that you can pull it apart like pages of a book. If you and your graduate students are willing to crawl over the weathered surfaces of the rotten clay, in order to pick up fruits and seeds as they weather out, you can find abundant fruits and seeds. We did just that on several collecting trips. Examples of such fruits are *Gordonia, Nyssa,* and *Crepetocarpon*, seen in Figures

Fig. 6. Claiborne Formation, middle Eocene, localities rich in fossil plants and from which the fossils discussed in this paper were collected. Modified from Potter and Dilcher 1980.

16, 17 and 21 as well as many other different fruits and seeds. Fruits and seeds are extremely diagnostic for the angiosperms and so they are important fossil plants to work with.

We also found structurally preserved flowers that are complete with pollen (Crepet et al. 1974a, 1975; Crepet & Dilcher 1977). We have over 20 localities in western Kentucky, western Tennessee, and Mississippi from which we have collected nearly 30,000 fossil plants (Fig. 6) over the past 40 years. The clay deposits are often covered over by river borne sands. The cross bedding of these sands is typical of river deposited sands. When we look at the outlines of these clay deposits we find they are crescent shaped, much like clay plugs in oxbow lakes on river flood plains. So I concluded that there were ancient river systems that, for several millions of years coursed through this area, cutting off oxbow lakes (Dilcher 1971). During flooding of the river fine grain clays were deposited in the oxbow lakes on the flood plain trapping large amounts of plant material. Rather than thinking these were deposited near the shore of an ocean as Berry proposed (Berry 1916, 1930), the ocean was probably south of the clay deposits by 50 miles or so (Potter & Dilcher 1980). Now when considering these fossil deposits, we're looking at a river system, not so different from the Mississippi River system accumulating clays in oxbow lakes. Each oxbow lake probably lasted no more than 1,000 to 1,500 years and thus each clay pit represents a short segment of time preserving floras that are dispersed through middle Eocene time.

NATURE OF THE FOSSIL RECORD

I will present some examples of the fossil material that has been found that demonstrates relationships to other fossils and living genera found in Central America, South America, Europe,

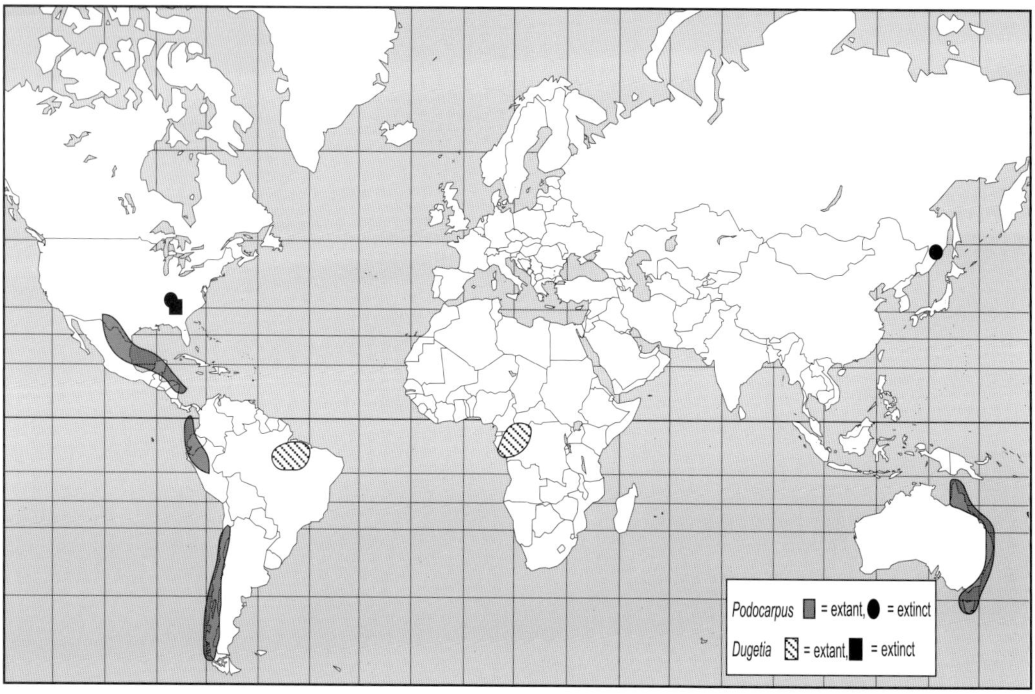

FIG. 7. Phytogeographic map of the distribution of *Podocarpus* and *Duguetia* showing their living and fossil occurrences.

Africa and Asia today. Many of these are genera that have their present, often disjunct, distributions explained by migrations along the borders of the Tethys Sea. Many of these fossil plants are tropical in nature, some have a distinct east-west distribution and some have their source in Gondwana.

There is a fossil gymnosperm that was identified as *Taxodium dubium* reported from the clay deposits (Berry 1916) (Fig 11). It was difficult to determine the affinities of *Taxodium dubium* by just looking at the leafy shoots and in fact it looked like a *Taxodium* shoot. However when I examined the cuticle of the leaves I found that there was a very peculiar thickened ring surrounding each stomata that is unique to the genus *Podocarpus* (Dilcher 1969). This character is never found in *Taxodium, Sequoia* or *Metasequoia*. *Podocarpus* is found as a Gondwana element and its current range extends only as far north today as Mexico. The genus is found in the higher elevations of South America and Central America today. Also, the genus has a range to New Zealand, Australia and New Caledonia as a Gondwana element. Krassilov (1974) reports *Podocarpus* fossils in eastern Siberia from the Tertiary. So it is clear that *Podocarpus* had an extensive history in the Northern Hemisphere that is not reflected by its current distribution (Fig. 7). The presence of *Podocarpus* in southeastern United States during the middle Eocene demonstrates the dispersal of plants from the tropical forests and Andean areas of South America northward through Central America and Mexico by Eocene times.

Berry (1916) described a large leaf from the clay sediments as a *Magnolia* leaf. Upon preparation of the cuticle it became obvious that this leaf was not *Magnolia*. As a result of comparisons with more tropical elements such as *Duguetia* in the Anonnaceae (Roth & Dilcher 1981; Roth 1981) we found that the fossil seemed to hold all of the form, venation and cuticular features of this living genus (Fig. 10). The current distribution of the genus *Duguetia* is in the tropical areas of the Brazilian amazon and in western Africa (Fig. 7). This disjunct distribution of a tropical genus is not difficult to understand if the distribution of this genus is considered to reflect the Pantropical Flora of the Late Cretaceous (Fig. 4).

Another fossil plant is a palm that I worked on together with Frank Potter and Charles Daghlian. We worked with the cuticle and the nature of the insertion of the leaflets and their

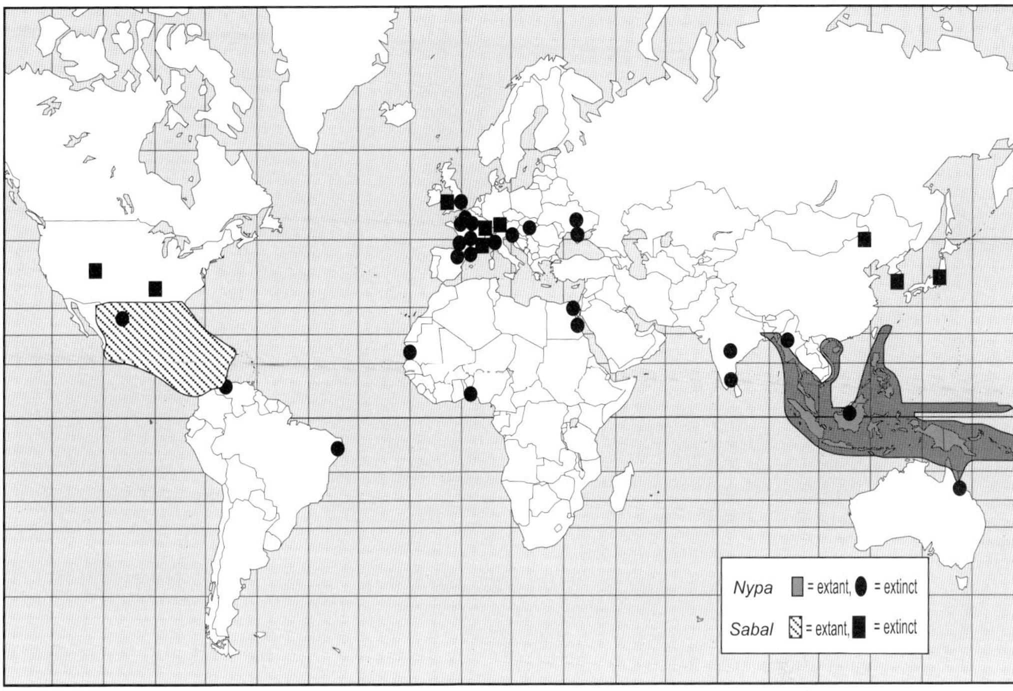

FIG. 8. Phytogeographic map of the distribution of *Nypa* and *Sabal* showing their living and fossil occurrences.

plications (Dilcher 1968) which matches closely with *Sabal*. Charles Daghlian later took over this project, adding new material from Texas Eocene sediments, and finished his Ph. D. dissertation working with these palms (Daghlian 1978). The palm is identical to the genus *Sabal* (Fig. 12), with 13 species living around the Caribbean today (Zona 1990). *Sabal* has had a wide distribution in the fossil record of Asia and Europe throughout the Tertiary (Li & Zheng 1995; Mai 1995). So *Sabal* seems to have had a much broader distribution in the past that it has today (Fig. 8).

Another palm found in Eocene sediments of Texas is the genus *Nypa* (Fig. 9). *Nypa* is known as a fossil from several areas in the Americas and Europe from pollen (Westgate & Gee 1990) and megafossils (Tralau 1964). Today *Nypa* is a mangrove palm that is restricted to tropical areas of Asia (Fig. 8). It has been found as fossil fruits associated with a vertebrate locality in the Eocene sediments exposed near Laredo, Texas (Dilcher, Westgate, Lott and Gee, work in preparation). It has no fossil record in temperate or cooler floras, and truly represents a tropical environment when it is found. Here is a plant whose distribution must have been a product of the tropical influence of the Tethys Sea (Figs. 4, 8). The presence of *Nypa* near Laredo, Texas, the numerous fossils from Europe and its present occurrence in tropical Asia demonstrates the east-west distribution of tropical elements along the Tethys Sea. Of course the fruits of this palm could easily float along the ocean currents of the Tethys Sea (Fig. 3) moving to shorelines with tropical mangrove forests.

The oak family, Fagaceae, has dominated the Northern Hemisphere and it is especially common in the southeast throughout the Tertiary in the pollen record (Elsik 1974; Elsik & Dilcher 1974; Frederiksen 1988) and the megafossil record (Dilcher & Mehrotra 1969; Jones & Dilcher 1988). The genus used by most paleobotanists around the Northern Hemisphere, *Dryophyllum*, was revised (Jones & Dilcher 1988; Jones et al. 1988), and a new name, *Berryophyllum* (Fig. 15) was introduced. Fruits associated with *Berryophyllum* in Europe, *Trigonobalanopsis*, were related to *Trigonobalanus* which lives today only in the tropics of Southeast Asia (Mai 1995). Nixon and Crepet (1989) found that some fossil reproductive material with affinities to the Fagaceae was most closely related to the living *Colombobalanus* found living today in the mountains of northern Colombia. So it appears that diverse forms similar to this tropical oak genus dispersed along

FIGS. 9–12. 9. *Nypa* (scale bar = 2.0 cm). 10. *Duguetia* (scale bar = 2.0 cm). 11. *Podocarpus* (scale bar = 1.0 cm). 12. *Sabal* (scale bar = 1.0 cm).

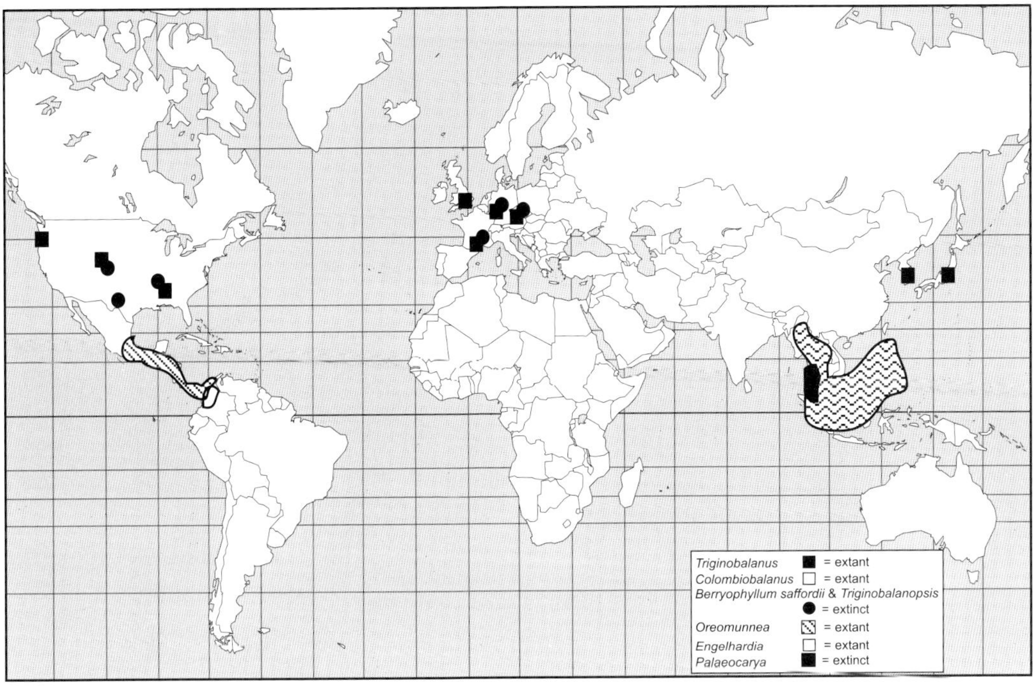

FIG. 13. Phytogeographic map of the distribution of *Trigonobalanus*, *Colombobalanus*, *Berryophyllum saffordii* and *Trigonobalanopsis*, and *Oreomunnea*, *Engelhardia*, and *Palaeocarya*.

the Tethys Sea and its remains are preserved in areas receiving sediment during the Eocene in Europe and southeastern United States. The disjunct distribution of *Trigonobalanus* and *Colombobalanus* today, are only the extreme ends of a previously extensive tropical range of these related taxa and their fossil relatives, which were influenced by the Tethys Sea (Fig. 13).

There were some members of the walnut family, Juglandaceae, living in the southeast between 38 and 50 million years ago (middle Eocene). This was during or just after the tremendous surge of co-evolution between plants and animals at which time winged fruits vs animal dispersed fruits were sorting themselves out for this family (Stone 1973). We began a study of the various organs of the Juglandaceae preserved in the southeast in the early 1970s (Crepet et al. 1974b; Potter et al. 1974; Dilcher & Crepet 1975). Fossil fruits (Dilcher et al. 1974), fossil catkins (Crepet et al. 1975) and fossil leaves (Dilcher & Manchester 1986) were studied that demonstrated the presence of some extinct genera such as *Palaeocarya* (Fig. 18). There is a good fossil record of this distinctive three-winged fruit known from the western part of North America, Europe and Japan (Manchester 1987). There are extant genera with similar winged fruits living today. The extant genera are *Oreomunnea* and *Alfaroa* (a non-winged genus) which lives in Central America and a closely related pair of genera recognized from tropical Asia, *Engelhardia* and *Alfaropsis* (Stone 1972; Iljinskaja 1994). *Oreomunnea* and *Alfaroa* live from 0° to 20° north of the equator and are confined to tropical environments. These extant genera are tropical elements within the Juglandaceae and their fossil record follows the Tethys Sea east and west (Fig. 13). Also, these fossil fruits are found mixed to the north with some more temperate elements during the Eocene in Oregon. These winged fruits may be found in both tropical and warm temperate climates of the past.

Nyssa has very durable endocarps that are frequently found as fossils in the Eocene clays (Dilcher & McQuade 1967). These fossil endocarps (Fig. 17) are sufficiently similar to the living genus *Nyssa,* that these fossils can easily be identified and related to it. However in spite of the fact that Berry (1916) identified leaves of *Nyssa* we have never been able to verify the presence of fossil leaves of this genus (Macklin & Dilcher 1980). *Nyssa* has a broad range today from the temperate and warm temperate areas of the United States, to tropical areas of Mexico, Panama

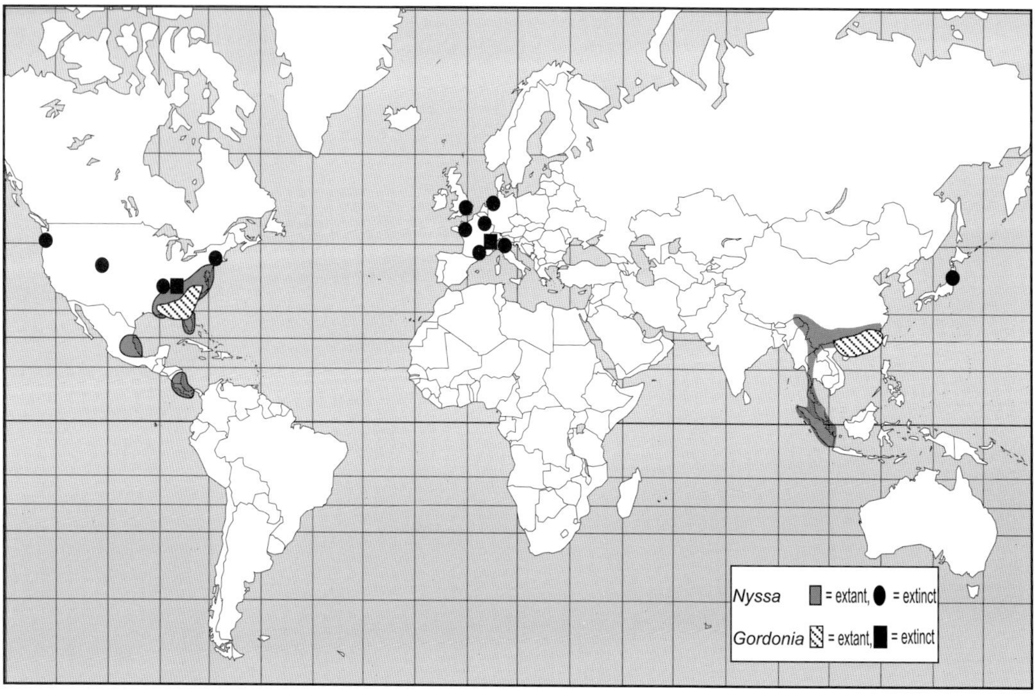

FIG. 14. Phytogeographic map of the distribution of *Nyssa* and *Gordonia* showing their living and fossil occurrences.

and Costa Rica (Hammel & Zamora 1990; Wen & Stuessy 1993), to temperate and warm temperate areas of China (Fig. 14). This disjunct distribution of living *Nyssa*, still extant on the extreme east and west ends of the old Tethys Sea, with numerous fossils found adjacent to the north of the seaway suggests that the genus could have dispersed along this tropical route and then spread north into more temperate areas. The alternative hypothesis is that *Nyssa* migrated across temperate land bridges to the far north and moved south into warm temperate and tropical areas (Manchester 1999).

The fruits of the Theaceae, especially *Gordonia*, are common in the clay deposits of western Kentucky and western Tennessee (Grote & Dilcher 1989, 1992). Through a lot of detailed work on the fruits and seeds of *Gordonia* and *Stewartia,* which are living today in southeastern United States and southeastern Asia (Fig. 14), we were able to determine that the fossil material (Fig. 16) has affinities with the living genus *Gordonia*. These genera are typical of warm temperate areas, for example *Gordonia* is common in the swamps of north Florida. It has beautiful white flowers, woody capsules and serrated leaves that are also abundant in the fossil record in North America and Europe. The fossil leaves are commonly known as the genus *Ternstroemites.* Fossil records of *Gordonia* and *Ternstroemites* or *Ternstroemia* are known from Europe and North America and make up some of the warm temperate elements of the common flora long the north coast of the Tethys Sea.

Another interesting extinct plant that we have collected as fruits, *Crepetocarpon* (Fig. 21), occurs in lignitic/clay sediments and has been compared with the living genus *Hippomane* (Lambert & Dilcher 1970; Dilcher & Manchester 1988). Fossils of extinct plants also related to this modern genus have been found in England (Chandler 1961) and Egypt (Chandler 1954). Extinct related fossils such as *Paleowetherellia* and *Wetherellia* have been reported from coastal Virginia and along the Gulf of Mexico coast in Mississippi (Mazer & Tiffney 1982; Call et al. 1993). Today these buoyant fruits often litter the beaches of many Caribbean Islands and Central America (Fig. 19). *Hippomane* is a small tree that grows along the strand line, quite close to the ocean and bears numerous fruits. This genus is currently distributed only along the tropical areas of the Caribbean Basin, Gulf of Mexico, and Pacific Ocean coasts of Central America and northern

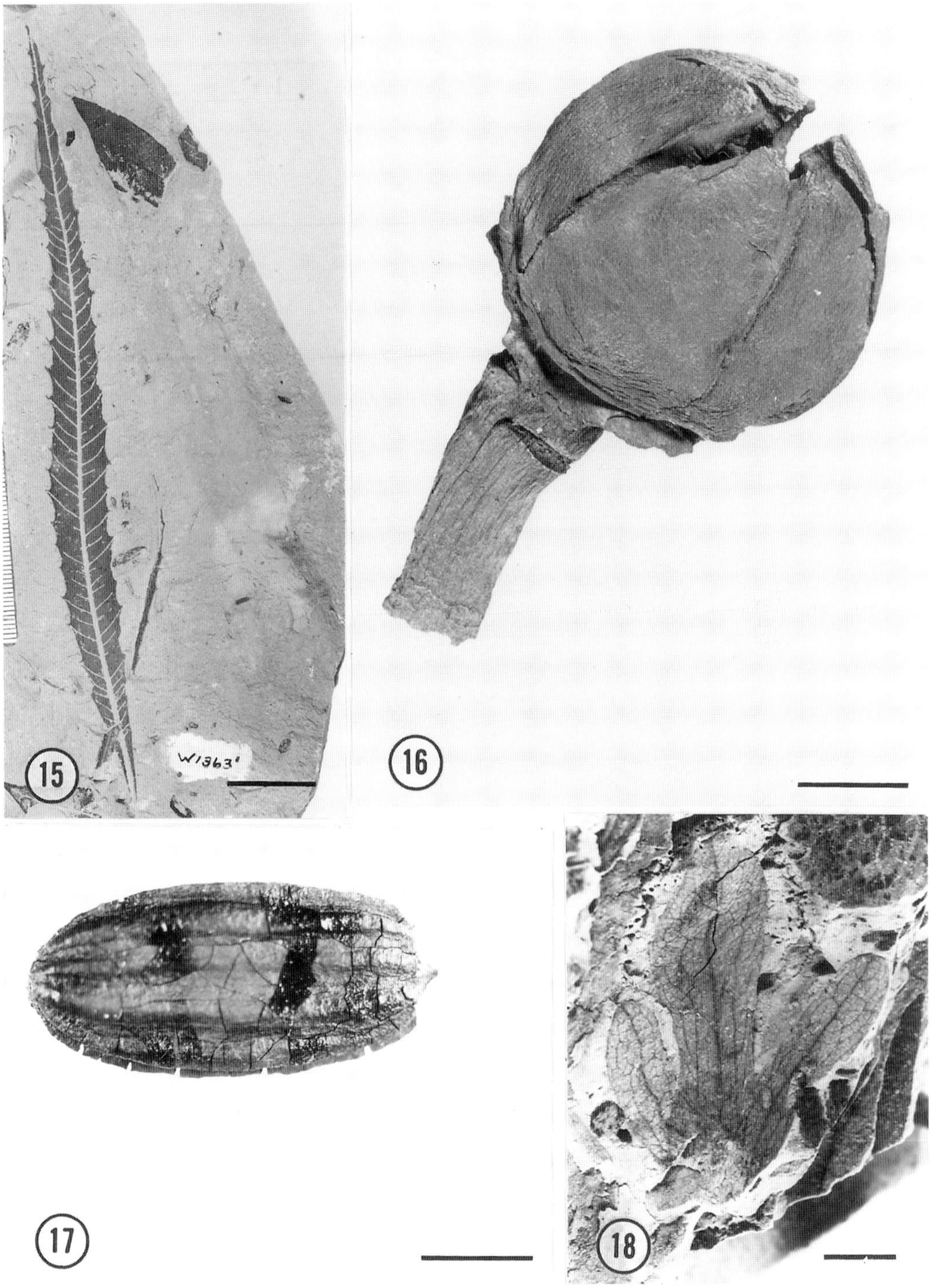

FIGS. 15–18. 15. *Berryophyllum saffordii* (scale bar = 2.0 cm). 16. *Gordonia* (scale bar = 0.5 cm). 17. *Nyssa* (scale bar = 1.5 cm). 18. *Palaeocarya* (scale bar = 0.5 cm).

South America (Fig. 19). The fossil record of closely related forms clearly demonstrates a Tethyan distribution and they probably were dispersed along the margins of the Tethys Sea. The distribution of this fruit, across what was a temperate North Atlantic region is doubtful and probably their dispersal took place along the warmer coasts of a warm Tethys Sea.

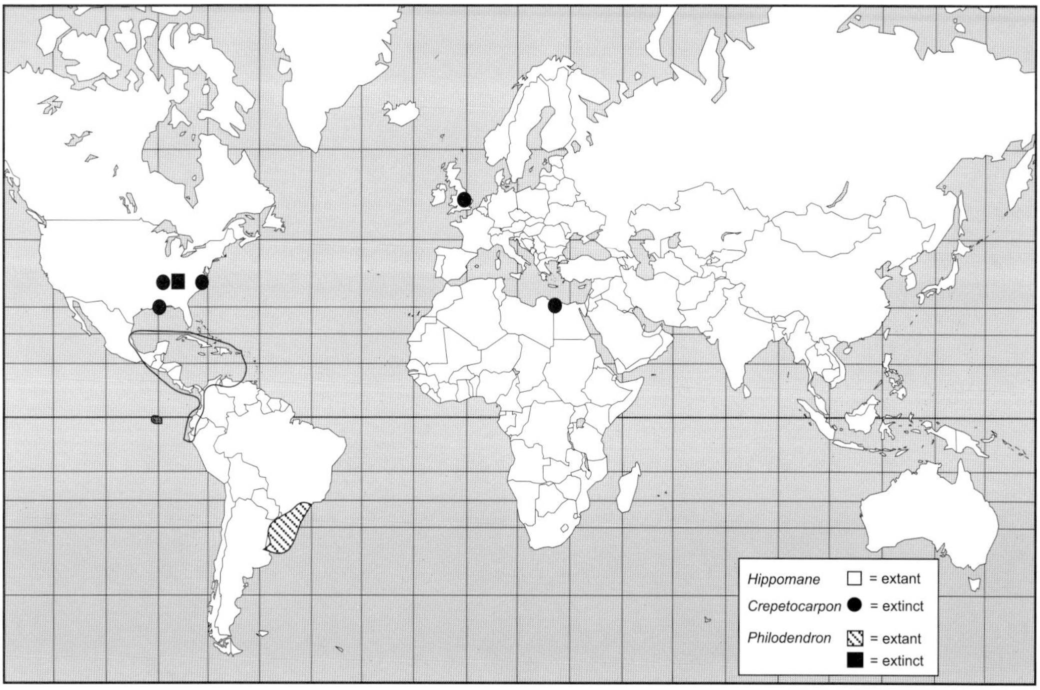

FIG. 19. Phytogeographic map of the distribution of *Hippomane*, *Crepetocarpon* and *Philodendron* showing their living and fossil occurrences.

Some genera, such as *Philodendron* (Fig. 22) in the fossil flora are restricted to the Americas and are found in Central America and South America today (Fig. 19). My wife, Katherine, and I collected a giant fossil *Philodendron* leaf (Dilcher & Daghlian 1977), one morning when we visited a clay pit in which they were actively mining the clay. The bucket of the dragline exposed a nearly complete leaf on a slab of clay over a meter square. The dragline operator lifted it out of the pit and laid it at our feet. We spent several hours trimming it down so we could lift it on top of our car (a VW bug) and carry it 300 miles home. This specimen is illustrated in Figure 22. *Philodendron* lives in warm temperate, subtropical, or tropical areas today. It can grow as a vine or a shrub in wet areas. Since this fossil was restricted to 2 clay pits out of over 20, and the associated flora was uncommon at the other pits, I think the presence of *Philodendron* represents a special environment that was wet, such as shallow standing water or a swampy area. The presence of this genus by the middle Eocene and its occurrence in both North America and South America clearly demonstrates the potential for dispersal between these two landmasses (Fig. 24).

Legumes are the most diverse species of fossils found in these Eocene clay localities. After collecting plant fossils from these clay pits for over 15 years it became obvious to me that legumes definitely were both abundant and diverse. Paula Weiss (1980) wrote her Masters thesis on the fossil mimosoids and then 10 years latter Patrick Herendeen wrote his Ph.D. dissertation on legume foliage and fruits in the flora (Herendeen 1990). It took nearly 30 years to collect sufficient material of leaflets, leaves, flowers and pods in order to provide the material necessary to undertake this research. We surveyed the entire diversity of legumes (Crepet & Dilcher 1977; Herendeen & Dilcher 1990a, 1990b, 1990c, 1991; Herendeen et al. 1992). It is interesting to note that areas with diverse legume species as a component of their floras usually are in the tropics. The Neotropics have the highest diversity, then Africa and finally Southeast Asia (Fig. 23) (Rundel 1989). While legumes are present in the Tertiary floras of Europe they are not as dominant as they are in the Tertiary floras of southeastern United States. These tropical legume elements probably dispersed along the landmasses forming the shores of the Tethys Sea resulting in the diversity observed today in the American and African tropics (Gentry 1993).

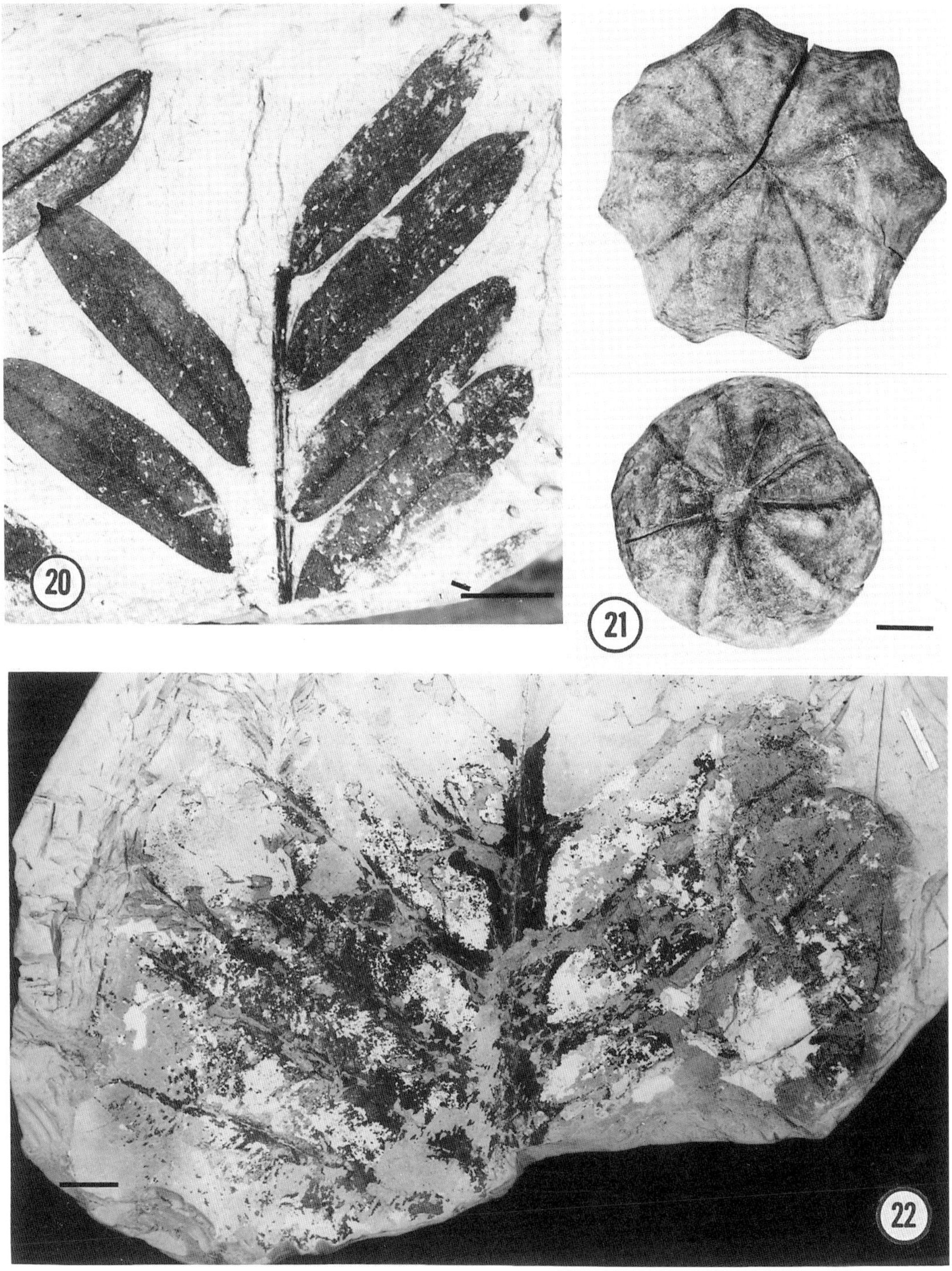

FIGS. 20–22. 20. Legume (scale bar = 0.5 cm). 21. *Crepetocarpon* (scale bar = 0.5 cm). 22. *Philodendron* (scale bar = 5.0 cm).

DISCUSSION

I have presented a few of the diverse elements in the Eocene Flora of southeastern United States. However there are certain distribution patterns and climatic limits of these elements that we can use to gain some insight into the distribution of vegetation in the tropical areas of the world. The Tethys Sea extended from the Asian tropics westward to the American tropics during the Upper Cretaceous and into the Tertiary. It was split asunder in the west by the opening of the

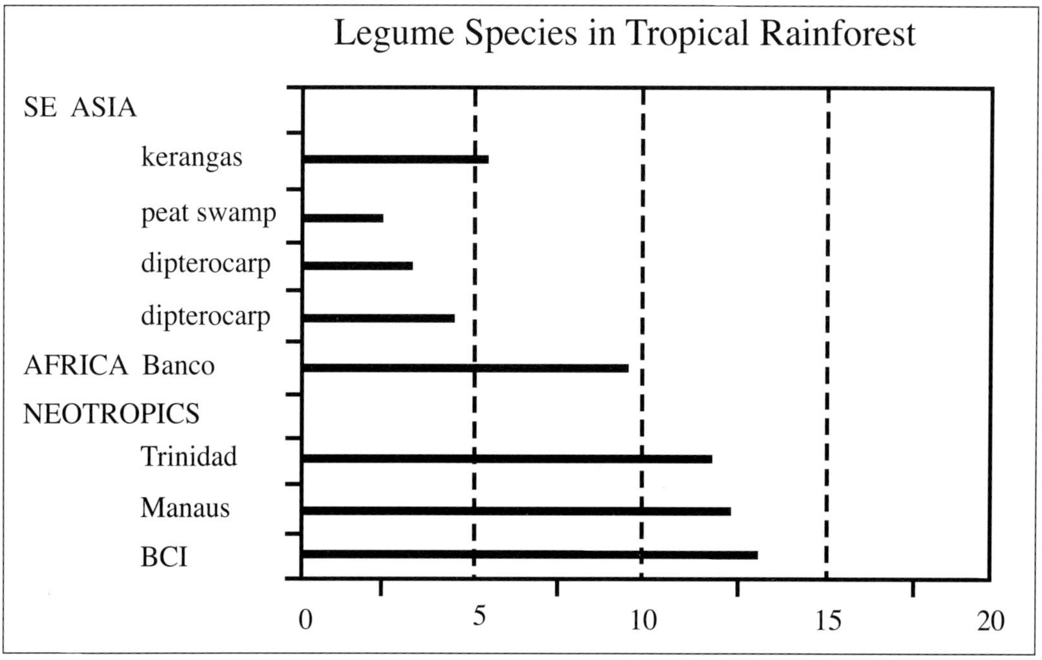

FIG. 23. Legume chart, showing the abundance of the legume species in Asia, Africa and South America. Modified from Rundel 1989.

Atlantic Ocean during the Upper Cretaceous and early Tertiary. The warm waters of the Tethys produced warm to tropical climates in the islands of Europe and North America and provided a seaway with westward trending currents and winds that could aid the migrations of plants. It was a broad expanse and mixture of deep and shallow seas so that it is not too difficult to imagine the fruits of *Nypa* and *Crepetocarpon* being carried around the margins of the Tethys Sea. Fossils of Eocene age sea cows are found scattered along its margins (Domning et al. 1982). The most complete fossil record that we have of Tertiary vegetation comes from northern areas of North America, Europe and Asia. Thus discussions of the relationships of floras between these areas most often are based upon elements that live in the cold temperate and temperate floras that are shared by these areas (Manchester 1999). However, there also are warm temperate, subtropical and tropical floral elements shared by these areas found in the fossil record that have related modern genera and living today in warm temperate, subtropical and tropical climates. These elements are more poorly documented and in need of careful research work. The tropical areas of the Mesozoic split apart during the Cretaceous at a time when the angiosperms were differentiating (Figs. 1, 3). The Cenomanian (Upper Cretaceous) was the time of the actual last land contact between Africa and South America (Maisey 1991). This was a time of extensive evolution of modern angiosperm families (Crepet 1999). The question remains, what was the potential for east and west migrations in tropical latitudes when the major continental lands were beginning to move apart? The Upper Cretaceous and Paleocene is the time we need to study and the borders of the Tethys Sea is the place we need to look for a record of the migration of tropical plants.

The Eocene floras from southeastern United States demonstrate clearly that there were migrations between South and North America as shown in Figure 24 (Taylor 1988; Jaramillo & Dilcher in press). We find *Philodendron, Duguetia* and *Berryophyllum* that all have affinities with Central and South American elements today. The presence of *Philodendron* and *Podocarpus* in southeastern United States illustrate north-south distributions that probably did not follow an east-west pattern as seen in other genera. *Podocarpus* follows a Gondwana distribution pattern while *Philodendron* is restricted to the new world and both probably migrated north from South America (Figs. 7, 19). Tropical elements that demonstrate an east-west connection are *Nypa, Berryophyllum, Duguetia* and *Crepetocarpon*. Warm temperate or subtropical elements that also have an east-west

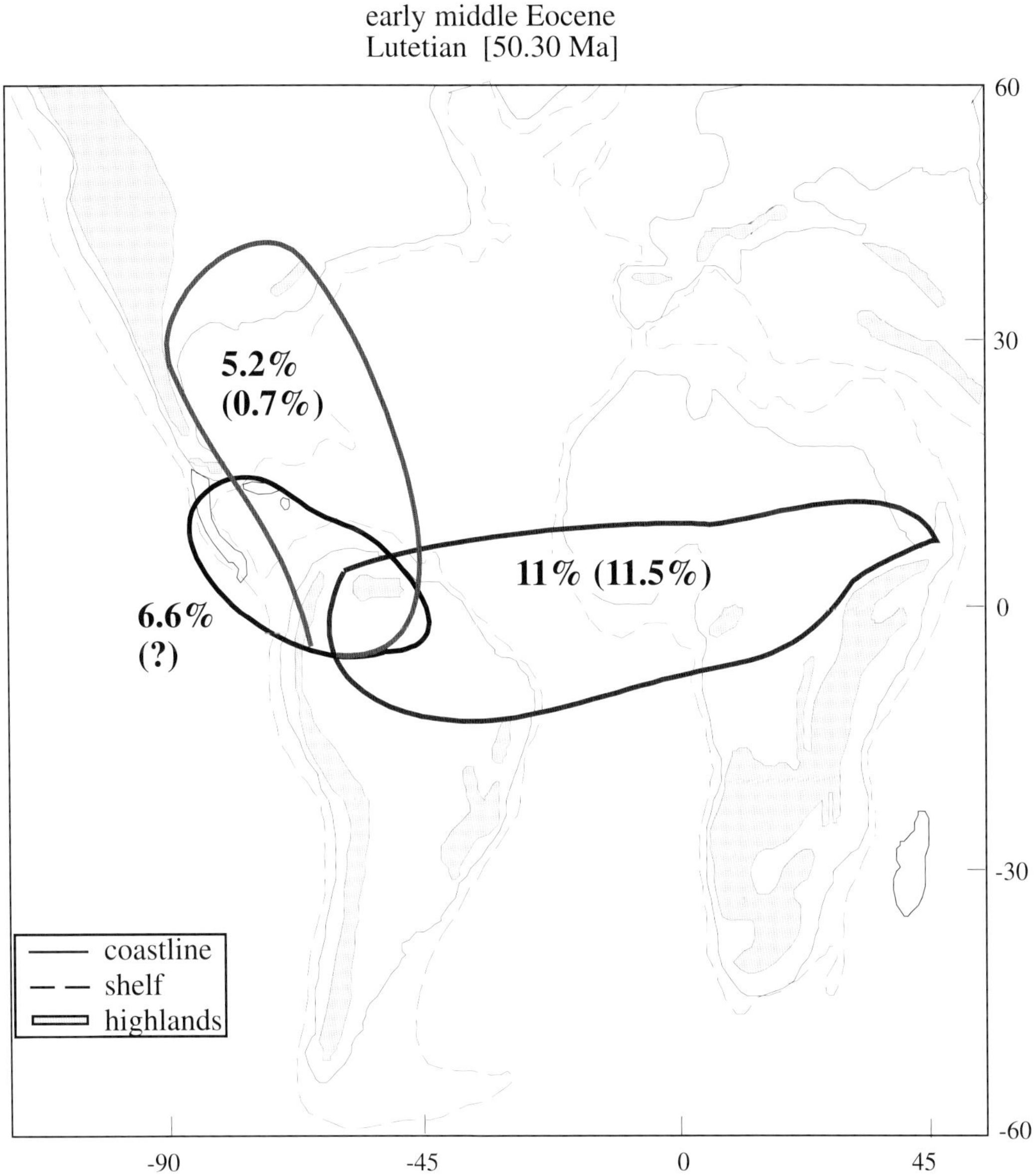

FIG. 24. Percentage of first occurrences of species shared between Africa and Colombia, Colombia and the Caribbean/ Central America, and Colombia and the Gulf Coast of North America during the Paleocene in bracts ( ) and the Eocene lacking bracts (after Jaramillo & Dilcher in press).

distribution include *Sabal*, *Palaeocarya*, *Nyssa* and *Gordonia*. The tropical to subtropical genera that migrated along Tethyan, east-west, low latitude routes probably also followed a north-south route to reach North America. Those north-south migrations probably took place at the close of the Paleocene or during the early Eocene as evidenced from the data presented in Figure 24.

The fossil record is more complete for the Laurasian Flora along the northern coastal areas of the Tethys (Mai 1995) than for the Pantropical Flora of tropical Asia, northern Africa and northern South America (Fig. 4). The same recovery pattern is true for Eocene sirenians that fed on the seagrass living along the shallow margins of the Tethys Sea (Domning et al. 1982). The distribution of these fossil sirenians links the old shorelines together with 11 finds in Europe and only 4 finds in north Africa. Africa is a difficult land to understand as a bridge for tropical plant distribution because so little is known about the fossil record there (Raven & Axelrod 1974) but the record is improving for Africa (Axelrod & Raven 1978; P. Herendeen pers. corr.). This is particularly true of the Cretaceous and early Tertiary floras of northern Africa (Schrank 1987, 1994; Awad 1994). Yet Africa stands between Asia and the Americas along the south shores of the Tethys Sea and must have been an important avenue in the dispersal of tropical elements of the world's vegetation. The relationships that Africa shares with South America are clear to see when we consider the data presented by Gentry (1993). He found that out of 93 families in a forest quadrat in Madagascar all families but 6 were also found in a quadrat sampled in South America. The land connections between Africa and South America continued into the Cenomanian and floristic exchange continued into the Paleocene and lower Eocene (Jaramillo & Dilcher in press). Certainly the chance for floristic exchange must have continued as angiosperms modernized during the Upper Cretaceous. The presence of *Duguetia* (Roth 1981; Roth & Dilcher 1981) living today in South America and Africa and also found in the Eocene sediments of southeastern United States supports the floristic relationships with tropical elements migrating between these three areas.

The current view is that angiosperm migrations around the Northern Hemisphere took place at high latitudes by way of the North Atlantic or Beringia at a time when the North Pole was warmer than it is today. It is uncertain and probably unlikely that these routes were frost free year around. These northern regions could not accommodate the subtropical and tropical plants that required frost free migration routes. The northern most occurrences of the fossils discussed in this paper are nearly always further south than 40–50° N. latitude. This requires migration when the warm Tethys Sea influenced the lower latitudes of the opening Atlantic Ocean. As presented here several of the tropical and subtropical elements that are found in the fossil flora of southeastern United States must have been distributed under the warm climatic influence of the Tethys Sea. This vegetation mixed with the floras to the north and as a result we have a unique and special mixture of tropical, subtropical, warm temperate and perhaps temperate fossil plants living together during the Tertiary in the southeast United States (Brown 1944; Gray 1960; Call & Dilcher 1992, 1997).

The floras living in the Northern Hemisphere are complex and the elements that make them up have long histories. Extinctions and missing fossil data dominate the record. The circumglobal east-west relationships of world floras are the rule. These world floras while showing east-west relationships extend from the North Pole to the South Pole as bands or Floral Provinces when pieced together with the help of the fossil record. Recent and past world-wide Floral Provinces relate to those in existence during the Upper Cretaceous (Axelrod 1970; Srivastava 1994) (Fig 4). The better we understand the history of these ancient Provinces the better we will be able to understand the evolution of world vegetation today.

ACKNOWLEDGEMENTS

I thank Sy Sohmer, and the staff at the Botanical Research Institute of Texas for the opportunity to present this paper. All of my students, friends, colleagues and family who have engaged in research on the fossil plants from southeastern United States or who have collected these plant fossils are gratefully acknowledged and thanked. This paper is the product of nearly 40 years of collections and work on these fossil floras. The author appreciates the helpful review of the manuscript by Scott Wing. The help of Terry Lott, who followed every aspect of the preparation of the plates and text of this paper, and Carlos Jaramillo who translated the resumen, provided helpful discussions and aided in the preparation of Figures 1–4 and 24 are acknowledged with thanks. Steven Manchester is thanked for his suggestions and corrections for this manuscript.

REFERENCES

ANDERSON, J.M. (ed.). 1999. Towards Gondwana Alive. Gondwana Alive Society, Pretoria.

AWAD, M.Z. 1994. Stratigraphic, palynological and paleoecological studies in the East-Central Sudan (Khartoum and Kosti Basins), Late Jurassic to Mid-Tertiary. Berliner geowiss Abh., A, 161:1–163.

AXELROD, D.I. 1970. Mesozoic paleogeography and early angiosperm history. Bot. Rev. 36:277-319.

________ and P.H. RAVEN. 1978. Late Cretaceous and Tertiary vegetation history of Africa. In: M.J.A. Werger, ed. Biogeography and Ecology of Southern Africa. W. Junk Publ., The Hague.

BASINGER, J.F. 1986. Our 'tropical' arctic. Canadian Geogr. 106:28–37.
BERRY, E.W. 1916. The lower Eocene floras of southeastern North America. U.S.G.S. Prof. Paper 91:1–481.
______________. 1922. Additions to the flora of the Wilcox Group. U.S.G.S. Prof. Paper 131A:1–21.
______________. 1924. The middle and upper Eocene floras of southeastern North America. U.S.G.S. Prof. Paper 92:206 p.
______________. 1930. Revision of the lower Eocene Wilcox flora of the southeastern States. U.S.G.S. Prof. Paper 156:1–196.
______________. 1941. Additions to the Wilcox flora from Kentucky and Texas. U.S.G.S. Prof. Paper 193E:83–99.
BROWN, R.W. 1944. Temperate species in the Eocene flora of the southeastern United States. J. Wash. Acad. Sci. 34:349–351.
CALL, V.B. and D.L. DILCHER. 1992. Investigations of Angiosperms from the Eocene of southeastern North America: Samaras of *Fraxinus wilcoxiana* Berry. Rev. Palaeobot. Palynol. 74:249–266.
______________ and ______________. 1997. The fossil record of *Eucommia* (Eucommiaceae) in North America. Amer. J. Bot. 84:798–814.
______________, S.R. MANCHESTER, and D.L. DILCHER. 1993. *Wetherellia* fruits and associated fossil plant remains from the Paleocene/Eocene Tuscahoma-Hachetigbee Interval, Meridian, Mississippi. Mississippi Geology 14:10–18.
CHANDLER, M.E.J. 1954. Some Upper Cretaceous and Eocene fruits from Egypt. Bull. British Mus. Nat. Hist., Geol. 2:149–187.
______________. 1961. Post-Ypresian plant remains from the Isle of Wight and Selsey Penninsula, Sussex. Bull. British Mus. Nat. Hist., Geol. 5:2–41.
CREPET, W.L. 1999. Early bloomers. Nat. Hist. 108:40–41.
______________ and D.L. DILCHER. 1977. Investigations of angiosperms from the Eocene of North America: a mimosoid inflorescence. Amer. J. Bot. 64:714–725.
______________, ______________, and F.W. POTTER. 1974a. Eocene angiosperm flowers. Science 185:781 782 and the cover photograph for that issue.
______________, ______________ and ______________. 1974b. Investigations of Angiosperms from the Eocene of southeastern North America: a new catkin from Tennessee. Amer. J. Bot. 61:S14. Abstract.
______________, ______________ and ______________. 1975. Investigations of angiosperms from the Eocene of southeastern North America: a Catkin with Junglandaceous affinities. Amer. J. Bot. 62:813–823.
______________ and K.C. NIXON. 1998. Fossil Clusiaceae from the Late Cretaceous (Turonian) of New Jersey and implications regarding the history of bee pollination. Amer. J. Bot. 85:1122–1133.
DAGHLIAN, C.P. 1978. Coryphoid palms from the lower and middle Eocene of southeastern North America. Palaeontographica 166B:44–82.
DILCHER, D.L. 1963. Cuticular analysis of Eocene leaves of *Octoea obtusifolia*. Amer. J. Bot. 50:1-8.
______________. 1968. Revision of Eocene palms from southeastern North America based upon cuticular analysis. Amer. J. Bot. 62:S639. Abstract.
______________. 1969. *Podocarpus* from the Eocene of North America. Science 164:299–301.
______________. 1971 (1973). A revision of the Eocene floras of southeastern North America. Paleobotanist 20:7–18.
______________. 1973. A paleoclimatic interpretation of the Eocene floras of southeastern North America. In: A. Graham, ed. Vegetation and vegetational history of northern Latin America. Elsevier Publ. Co., Amsterdam. Chapter 2, pp. 39–59.
______________. 1974. Approaches to the identification of fossil leaf remains. Bot. Rev. 40:1–157.
______________ and W.L. CREPET. 1975. Juglandaceous leaflet remains. Botanical Society Meetings, Oregon State University, Corvallis. Abstract, pp. 19.
______________ and C.P. DAGHLIAN. 1977. Investigations of angiosperms from the Eocene of southeastern North America: *Philodendron* leaf remains. Amer. J. Bot. 64:526–534.
______________ and S.R. MANCHESTER. 1986. Investigations of Angiosperms from the Eocene of North America: Leaves of the Engelhardieae (Juglandaceae). Bot. Gaz. 147:189–199.
______________ and ______________. 1988. Investigations of the angiosperms from the Eocene of North America: A fruit belonging to Euphorbiaceae. Tertiary Res. 9:45–68.
______________ and J.F. MCQUADE. 1967. A morphological study of *Nyssa* endocarps from Eocene deposits in western Tennessee. Torrey Bot. Club Bull. 94:35–40.
______________ and B. MEHROTRA. 1969. A study of leaf compressions of *Knightiophyllum* from Eocene deposits of southeastern North America. Amer. J. Bot. 56:936–943.
______________, F.W. POTTER, and W.L. CREPET. 1974. Investigations of angiosperms from the Eocene of North America: Juglandaceous winged fruits. Amer. J. Bot. 63:532–544.
DOMNING, D.P., G.S. MORGAN, and C.E. RAY. 1982. North American Eocene sea cows (Mammalia: Sirenia). Smith. Contr. Paleob. 52:1–69.
ELSIK, W.C. 1974. Characteristic Eocene palynomorphs in the Gulf Coast, U.S.A. Palaeontographica 149B:90–111.
______________ and D.L. DILCHER. 1974. Palynology and age of clays exposed in Lawrence Clay Pit, Henry County, Tennessee. Palaeontographica 146B:66–87.

FREDERIKSEN, N.O. 1988. Sporomorph biostratigraphy, floral changes, and paleoclimatology, Eocene and earliest Oligocene of the eastern Gulf Coast. U.S.G.S. Prof. Paper 1448:1–68.

GANDOLFO, M.A., K.C. NIXON, and W.L. CREPET. 1998. *Tylerianthus crossmanensis* Gen. et sp. nov. (aff. Hydrangeaceae) from the upper Cretaceous of New Jersey. Amer. J. Bot. 85:376–386.

GENTRY, A.H. 1993. Diversity and floristic composition of lowland tropical forest in Africa and South America. In: P. Goldblatt, ed. Biological relationships between Africa and South America. Yale University Press, New Haven.

GRAHAM, A. 1999. Late Cretaceous and Cenozoic History of North American vegetation. Oxford University Press, New York.

GRAY, J. 1960. Temperate pollen genera in the Eocene (Claiborne) Flora, Alabama. Science 132:808–810.

GROTE, P.J. and D.L. DILCHER. 1989. Investigations of angiosperms from the Eocene of North America: A new genus of Theaceae based on fruit and seed remains. Bot. Gaz. 150:190–206.

——— and ———. 1992. Fruits and seeds of Tribe Gordonieae (Theaceae) from the Eocene of North America. Amer. J. Bot. 79:744–753.

HAMMEL, B.E. and N.A. ZAMORA. 1990. *Nyssa talamancana* (Cornaceae), an addition to the remnant Laurasian Tertiary flora of southern Central America. Brittonia 42:165–170.

Herendeen, P.S. 1990. Fossil history of the Leguminosae from the Eocene of southeastern North America. PhD, Biology, Indiana University.

———, W.L. CREPET, and D.L. DILCHER. 1992. The fossil history of the Leguminosae: Phylogenetic and biogeographic implications. In: P.S. Herendeen and D.L. Dilcher, eds. Advances in legume systematics, Part 4. The fossil record. Pp. 303–316.

——— and D.L. DILCHER. 1990a. Fossil mimosoid legumes from the Eocene and Oligocene of southeastern North America. Rev. Palaeobot. Palynol. 62:339–361.

——— and ———. 1990b. *Diplotropis* (Leguminosae, Papilionoideae) from the middle Eocene of southeastern North America. Syst. Bot. 15: 526–533.

——— and ———. 1990c. Reproductive and vegetative evidence for the occurrence of *Crudia* (Leguminosae, Caesalpinioideae) in the Eocene of southeastern North America. Bot. Gaz. 151:402–413.

——— and ———. 1991. *Caesalpinia* subgenus Mezoneuron (Leguminosae, Caesalpinioideae) from the Tertiary of North America. Amer. J. Bot. 78:1–12.

ILJINSKAJA, I. 1994. Juglandaceae. In: L. Budantsev, ed. Fossil flowering plants of Russia and adjacent states. Volume 3 Leitneriaceae - Juglandaceae. Fedorov Printing House, Russia. Pp 37–106.

JARAMILLO, C.A. and D.L. DILCHER (In press). Middle Paleogene palynology of Central Colombia, South America: A study of pollen and spores from tropical latitudes. Paleontographica B.

JONES, J.H. and D.L. DILCHER. 1988. A study of the "Dryophyllum" leaf forms from the Paleocene of southeastern North America. Palaeontographica 208B:53–80.

———, S.R. MANCHESTER and D.L. DILCHER. 1988. *Dryophyllum* Debey ex Saporta, Juglandaceous not Fagaceous. Rev. Palaeobot. Palynol. 56:205–211.

KRASSILOV, V.A. 1974. *Podocarpus* from the Upper Cretaceous of eastern Asia and its bearing on the theory of conifer evolution. Palaeontology 17:365–370.

Lambert, N.E. and D.L. DILCHER. 1970. Eocene Euphorbiaceous fruits. Indiana Acad. Sci. 79:375. Abstract.

LI, HAOMIN and Y. Zheng. 1995. Paleogene Floras. In: Li Xingxue, ed. Fossil floras of China through the geological ages. Guangdong Science and Technology Press, Guangzhou, China, pp. 455–505.

MACKLIN, W.D. and D.L. DILCHER. 1980. Leaf morphology of *Nyssa*. A description of the modern species. Proc. Ind. Acad. Sci. 89:95–96. Abstract.

MAI, D.H. 1995. Tertiäre Vegetatonsgeschichte Europas. Gustav Fischer, Jena.

MAISEY, J.G. 1991. Santana Fossils: An illustrated atlas. T.F.H. publications, Inc. Neptune, N.J.

MANCHESTER, S.R. 1987. The fossil history of the Juglandaceae. Monogr. Missouri Bot. Gard. 21:1–137.

———. 1999. Biogeographical relationships of North American Tertiary Floras. Ann. Missouri Bot. Gard. 86: 472–522.

MAZER, S.J. and B.H. TIFFNEY. 1982. Fruits of *Wetherellia* and *Palaeowetherellia* (?Euphorbiaceae) from Eocene sediments in Virginia and Maryland. Brittonia 34:300–333.

MURRAY, G.E. 1961. Geology of the Atlantic and Gulf Coastal province of North America. New York: Harper.

NIXON, K.C. and W.L. CREPET. 1989. *Trigonobalanus* (Fagaceae): Taxonomic status and phylogenetic relationships. Amer. J. Bot. 76:828–841.

POTTER, F. and D.L. DILCHER. 1980. Biostratigraphic analysis of middle Eocene floras of western Kentucky and Tennessee. In: D.L. Dilcher and T. Taylor, eds. Biostratigraphy of fossil plants: Successional and paleoecological analysis. Dowden, Hutchinson and Ross, Publishers. Chapter 8, pp. 211–225.

———, ——— and W.L. CREPET. 1974. Investigations of Angiosperms from the Eocene of southeastern North America: winged fruits of the Juglandaceae. Amer. J. Bot. 61:S18. Abstract.

RAVEN, P. and D.I. AXELROD. 1974. Angiosperm biogeography and past continental movements. Ann. Missouri Bot. Gard. 61:539–673.

ROTH, J. 1981. Epidermal studies in the Annonaceae and related families. Ph.D., Indiana University.

——— and D.L. DILCHER. 1981. Foliar trichomes of the Annonaceae. Amer. J. Bot. Misc. Series. 160:8. Abstract.

RUNDEL, P.W. 1989. Ecological success in relation to plant form and function in the woody legumes. In: C.H. Stirton and J.L. Zarucchi, eds. Advances in legume biology. Monogr. Syst. Bot. Missouri Bot. Gard. 29:377–398.

SCHRANK, E. 1987. Palaeozoic and Mesozoic palynomorphs from northeast Africa (Egypt and Sudan) with special reference to Late Cretaceous pollen and dinoflagellates. Berliner geowiss. Abh., A, 75:249–310

SCHRANK, E. 1994. Palynology of the Yesomma Formation in northern Somalia: A study of pollen, spores and associated phytoplankton from the Late Cretaceous Palmae Province. Palaeontographica, Abt. B, 231:63–112.

SRIVASTAVA, S.K. 1981. Evolution of Upper Cretaceous phytogeoprovinces and their pollen flora. Rev. Palaeobot. Palynol. 35:155–173.

______________. 1994. Evolution of Cretaceous phytogeoprovinces, continents and climates. Rev. Palaeobot. Palynol. 82: 197–224.

STONE, D.E.. 1972. New World Juglandaceae III: A new perspective of the tropical members with winged fruits. Ann. Missourii Bot. Gard. 59:297–321.

______________. 1973. Patterns in the evolution of amentiferous fruits. Brittonia 25:371–384.

TAYLOR, D.W. 1988. Paleobiogeographic relationships of the Paleogene flora from the southeastern U.S.A.: implications for West Gondwanaland affinities. Palaeogeogr., Palaeoclimatol., Palaeoecol. 66:265–275.

TIFFNEY, B.H. 1985. The Eocene North Atlantic land bridge: its importance in Tertiary and modern phytogeography of the Northern Hemisphere. J. Arnold Arbor. 66:243–273.

TRALAU, H. 1963. Asiatic dicotyledonous affinities in the Cainozoic flora of Europe. Kgl. Svensk Vet.akad. Handl. 9:1–87.

______________. 1964. The genus *Nypa* van Wurmb. Kgl. Svenska Vet.akad. Handl. 10:5–29.

WEISS, J. 1980. Fossil mimosoid leaflets from the middle Eocene (Claiborne Formation), Graves County, Kentucky. M.S., Geology, Indiana University.

WEN, J. and T.F. STUESSY. 1993. The phylogeny and biogeography of *Nyssa* (Cornaceae). Syst. Bot. 18:68–79.

WESTGATE, J.W. and C.T. GEE. 1990. Paleoecology of a middle Eocene mangrove biota (vertebrates, plants, and invertebrates) from southwest Texas. Palaeogeogr., Palaeoclimat., Palaeoecol. 78:163–177.

ZONA, S. 1990. A monograph of *Sabal* (Arecaceae: Coryphoideae). Aliso 12:583–666.

THE AQUATIC VASCULAR FLORA OF THE SOUTHEASTERN UNITED STATES: ENDEMISM AND ORIGINS

ROBERT R. HAYNES

Biodiversity and Systematics
Department of Biological Sciences
The University of Alabama
Tuscaloosa, AL 35487-0345, U.S.A.

The definition of aquatic vascular plants varies among botanists. Some botanists, for example, broaden the definition to include those species that occur in wetland situations, whereas other botanists limit the definition to include only those species of plants that grow and reproduce throughout the year in standing or flowing water. The latter is the definition used here. These species are sometimes considered to be "true aquatic plants." Certainly, many members of aquatic flowering plants occasionally are stranded during times of low water, but they flower or fruit under normal conditions when they are growing in an aquatic environment.

Habitats in which aquatic vascular plants occur are as varied as their definitions. These habitats are lakes, reservoirs, rivers, small streams, ephemeral pools, and swamps. Certainly the latter two habitats are often dry during the summers, but the aquatic vascular plants that inhabit these areas usually go into a dormant stage during these times of drought. Habitats omitted for "true aquatic vascular plants" are flood plains of rivers, seeps, pine flat woods, and wet ditches. Floodplains may be inundated during high-water periods, but that is only a temporary condition and is not considered a habitat for true aquatic plants.

To study aquatic vascular plants, one must go into the water with the plants. The easiest method, and the one that I utilize mostly, is to wade, often waist deep or more. One may use rubber boots, including chest or hip waders; however, during warmer seasons and in areas where health is not a problem, I just wade in sneakers and jeans. Additional methods include dragging hooks or racks from shore or a boat or, of course, grabbing the plants by hand from a boat. Finally, one can certainly use SCUBA (Self Contained Breathing Apparatus) and swim in among the plants. Regardless of the method chosen, one must get in with the plants.

OVERVIEW OF AQUATIC VASCULAR PLANTS IN THE SOUTHEAST

The southeastern United States as here defined follows essentially that outlined in Cronquist (1980). It encompasses an area from Delaware and Maryland west to Kentucky, Tennessee, and Louisiana and all states to the south and east. I have chosen to omit Arkansas, which is included in Cronquist (1980) and east Texas, which is not, since both states have a large percentage of their flora more similar to those further north or west than do they to other states of the Southeast.

Aquatic vascular plants are distributed throughout much of the plant kingdom. In the southeastern United States, these species occur in five families of Pteridophyta and all of the subclasses of the Magnoliophyta (Table I). None are known from the southeast in the Pinophyta. *Taxodium distichum* (L.) Richard, however, does occur in swamps, but it also can exist and reproduce very well out of water.

As indicated in Table 1, 263 species of aquatic vascular plants in 62 families and 112 genera occur in the southeastern United States. Among these 263 species, 33 are endemics (Table 2), 91 are restricted to North America north of Mexico but occur in some state or province outside of the southeast, and 139 species are in areas outside of North America north of Mexico in addition to North America.

Aquatic vascular plant species that are endemic to the Southeast are distributed throughout the plant kingdom, seven pteridophytes, 16 dicots, and 10 monocots. The subclass with the highest degree of endemics is the Alismatidae with eight. The Alismatidae, of course, is the only subclass that is composed entirely of aquatic or wetland species. The high degree of endemism

SIDA, Bot. Misc. 18: 23–28. 2000

among the pteridophytes is in *Isoëtes*, which represents all of the pteridophhyte endemics. The majority of these *Isoëtes* endemics are species restricted to ephemeral pools of piedmont and cumberland plateau provinces.

The 33 endemic species are concentrated in the four southeastern most states, Alabama (18 species), Florida (19 species), Georgia (19 species), and South Carolina (12 species). Seven of Georgia's endemic species are *Isoëtes*, three of which are restricted to Georgia and an additional one is known from one population in South Carolina (Taylor et al. 1993). As the distance from these four states increases, especially to the north, the number of endemic species decreases. North Carolina has nine endemic species, Louisiana and Mississippi each have five, Virginia three, Tennessee one, and Kentucky and West Virginia none. It is difficult to speculate as to the cause of such a high degree of endemism in these three states other that possibly a combination of more diverse habitats and more tropical environment.

In addition to native species, the aquatic flora of the Southeast consists of 26 non-indigenous species (Table 3), including 16 dicots and 10 monocots. No confirmed naturalized aquatic ferns are in the southeast (Lloyd 1993; Nauman 1993; Lumpkin 1993), although the name *Salvinia auriculata* Aublet has been misapplied (Nauman 1993).

These non-indigenous species are pest plants of water ways, plants of rice fields, and tropical exotics. Among those species known to become pests in water ways, for example, is *Eichhornia crassipes* (Martius) Solms, the water hyacinth, generally considered to be the world's most serious aquatic weed according to Rosatti (1987). He continuous to note that floating mats of the species frequently obstruct navigation, impede drainage, contribute to eutrophication, and cause wastage of impounded water by displacement and transpiration.

Reasons for the invasion into the United States of aquatic pest plants are certainly varied, but many of them, *Eichhornia crassipes*, *Hydrocleys nymphoides*, *Limnophila sessiliflora*, *L. indica*, *Nymphaea ampla*, *N. elegans, N. capensis*, *N. lotus*, and *Egeria densa*, were probably a result of the aquarium trade, as described by Rosatti (1987) for *Eichhornia crassipes*. Others, *Ottelia alismoides*, *Blyxa aubertii*, and *Sagittaria guayanensis*, appeared first in or near rice fields in Louisiana and possibly were brought via seed from their native ranges.

A story known to only a few people concerns the first collections of *Blyxa aubertii* for North America. While a graduate student in south Louisiana, I was preparing a treatment of *Potamogeton* (Potamogetonaceae) for Louisiana (Haynes 1969) and visited the LSU herbarium to study the pondweeds there. Dr. C. A. Brown, then curator at LSU, was telling me about an unknown aquatic that he has recently collected in rice fields near Crowely, which is just west of Lafayette. It was so recently collected, in fact, that it was still in the plant press. He opened the press for me to examine, and I had no idea what it was. I returned to Lafayette and forgot about the specimen. A few days later, J.W. Thieret, D. Dike, and I were in Lacassine National Wildlife Refuge, near the town of Cameron, when I grabbed an unusual plant. After returning to Lafayette, I determined it to be *Blyxa aubertii* Richard, and we reported it new to North America (Thieret et al. 1969). After Dr. Brown's death years later, I realized that the specimen he had opened his press for me to examine was indeed *Blyxa aubertii*, a species he never received credit for having been the first to discover and to collect in North America.

Exotic species have been brought to the Southeast from many places, but predominantly from tropical areas (Table 4). Thirteen species are from tropical America, by far the major source of these aquatic species for the Southeast. The remaining ones are from Europe, temperate Asia, tropical Asia, and Africa.

GENERA OF SOUTHEASTERN UNITED STATES AQUATIC PLANTS

Table 5 lists the species of true aquatic species, in accordance with my above definition of these plants, known to be growing without cultivation in the Southeast. The Pteridophyta, Magnoliidae, and Hamamelidae are taken from the appropriate Flora of North America treatments, the Alismatidae from my manuscripts submitted to FNA (some co-authored by C.B. Hellquist), and the remainder from Godfrey and Wooten (1979, 1981) unless otherwise indicated.

TABLE 1. Distribution in the plant kingdom of southeastern United States aquatic vascular plants.

Superfamilial taxon	No. families	No. genera	No. species
Pteridophyta	5	5	23
Magnoliidae	6	8	23
Caryophyllidae	3	4	6
Dilleniidae	6	7	13
Rosidae	11	20	38
Asteridae	10	16	30
Alismatidae	8	21	63
Arecidae	2	8	19
Commelinidae	7	17	35
Zingiberidae	1	1	2
Lilidae	3	5	11
Total taxa	62	112	263

Table 2. Aquatic vascular plant species endemic to southeastern United States.

Isoëtes georgiana	*Arenaria alabamensis*	*Utricularia floridana*
Isoëtes flaccida	*Hibiscus grandiflorus*	*Potamogeton floridanus*
Isoëtes louisianensis	*Hibiscus coccineus*	*Najas filifolia*
Isoëtes boomii	*Stillingia aquatica*	*Echinodorus floridanus*
Isoëtes tegetiformans	*Ludwigia arcuata*	*Sagittaria isoetiformis*
Isoëtesmelanospora	*Myriophyllum laxum*	*Sagittaria secundifolia*
Isoëtes virginica	*Ptilimnium fluviatile*	*Sagittaria kurziana*
Nuphar ulvacea	*Oxypolis greenmanii*	*Sagittaria fasciculata*
Nuphar sagittifolia	*Crataegus aestivalis*	*Halophila johnsonii*
Nuphar orbiculata	*Hydrolea quadrivalvis*	*Zizania aquatica*
Arenaria glabra	*Amphianthus pusillus*	*Hymenocallis coronaria*

TABLE 3. List of non-indigenous species of aquatic vascular plants in the southeastern United States.

	Dicots	
Nelumbo nucifera	*Ludwigia peruviana*	*Caperonia palustris*
Nymphaea ampla	*Ludwigia peploides*	*Nymphoides peltata*
Nymphaea elegans	*Ludwigia uruguayensis*	*Limnophila sessiliflora*
Nymphaea capensis	*Myriophyllum aquaticum*	*Limnophila indica*
Nymhaea lotus	*Myriophyllum spicatum*	
Alternanthera philoxeroides	*Caperonia castaneifolia*	
	Monocots	
Sagittaria guayanensis	*Hydrocleys nymphoides*	*Hydrilla verticillata*
Potamogeton crispus	*Ottelia alismoides*	*Eichhornia crassipes*
Najas wrightiana	*Blyxa aubertii*	
Typha angustifolia	*Egeria densa*	

TABLE 4. Regions of origin for non-indigenous species of aquatic vascular plants.

Species	Region of origin	Type pest
Alternanthera philoxeroides	Latin America	water ways
Blyxa aubertii	Southeast Asia	rice fields, water ways
Caperonia castaneifolia	Latin America	water ways
Caperonia palustris	Latin America	water ways
Egeria densa	Latin America	water ways
Eichhornia crassipes	Latin America	water ways
Hydrilla verticillata	Europe/Asia	water ways
Hydrocleys nymphoides	Latin America	tropical exotic
Limnophila indica	Southeast Asia	water ways
Limnophila sessiliflora	Southeast Asia	water ways

TABLE 4. continued

Species	Region of origin	Type pest
Ludwigia peploides	Latin America	water ways
Ludwigia peruviana	Latin America	water ways
Myriophyllum aquaticum	Latin America	water ways
Myriophyllum spicatum	Europe/Asia	water ways
Najas wrightiana	Latin America	water ways
Nelumbo nucifera	Europe/Asia	water ways
Nymphaea ampla	Latin America	tropical exotic
Nymphaea capensis	Africa	tropical exotic
Nymphaea elegans	Latin America	tropical exotic
Nymphaea lotus	Africa	tropical exotic
Nymphoides peltata	Europe/Asia	water ways
Ottelia alismoides	Southeast Asia	rice fields, water ways
Potamogeton crispus	Europe/Asia	water ways
Sagittaria guayanensis	Latin America	rice fields, water ways
Typha angustifolia	Europe/Asia	water ways

TABLE 5. Species of aquatic vascular plants known to occur in the southeastern United States.

PTERIDOPHYTA

ISOËTACEAE
Isoëtes lacustris L.
Isoëtes georgiana Luebke
Isoëtes caroliniana (Eaton) Luebke
Isoëtes flaccida Braun
Isoëtes engelmannii Braun
Isoëtes riparia Engelm. ex Braun
Isoëtes louisianensis Thieret
Isoëtes boomii Luebke
Isoëtes tegetiformans Rury
Isoëtes melanospora Engelm.
Isoëtes butleri Engelm.
Isoëtes melanopoda Gay & Durieu
Isoëtes virginica Pfeiffer

PARKERIACEAE
Ceratopteris thalictroides (L.) Brongniart
Ceratopteris pteridoides (Hook.) Hieronymus
Ceratopteris richardii Brongniart

MARSILEACEAE
Marsilea quadrifolia L.
Marsilea ancylopoda Braun
Marsilea macropoda Engelm. ex Braun
Marsilea vestita Hook. & Greville

SALVINIACEAE
Salvinia minima Baker

AZOLLACEAE
Azolla caroliniana Willd.

MAGNOLIOPHYTA

Magnoliopsida

Magnoliidae

SAURURACEAE
Saururus cernuus L.

NELUMBONACEAE
Nelumbo lutea Willd.
Nelumbo nucifera Gaertn.

NYMPHAEACEAE
Nuphar advena (Aiton) Aiton
Nuphar ulvacea (Miller & Standley) Standley
Nuphar orbiculata (Small) Standley
Nuphar sagittifolia (Walter) Pursh
Nymphaea ampla (Salisbury) DC.
Nymphaea elegans Hook.
Nymphaea capensis Thunb.
Nymphaea jamesoniana Planchon
Nymphaea lotus L.
Nymphaea mexicana Zuccarini
Nymphaea odorata Aiton

CABOMBACEAE
Cabomba caroliniana Gray
Brasenia schreberi Gmelin

CERATOPHYLLACEAE
Ceratophyllum demersum L.
Ceratophyllum muricatum Chamisso
Ceratophyllum echinatum Gray

RANUNCULACEAE
Ranunculus flabellaris Raf.
Ranunculus sceleratus L.
Ranunculus aquatilis L.
Ranunculus laxicaulis Darby

Caryophyllidae

CARYOPHYLLACEAE
Arenaria alabamensis (McCormick, Bozeman, & Spongberg) Wyatt
Arenaria glabra Michx.

POLYGONACEAE
Rumex verticillatus L.
Polygonum amphibium L.
Polygonum hydropiperoides Michx.

PLUMBAGINACEAE
Limonium carolinianum (Walter) Britton

Dilleniidae

ELATINACEAE (after Tucker 1986; Haynes 1996)
Elatine americana (Pursh) Arnott
Elatine brachysperma Gray
Elatine triandra Schkur

PRIMULACEAE
Hottonia inflata Elliott

CLUSIACEAE
Hypericum mutilum L.
Hypericum galioides Lam.

MALVACEAE
Hibiscus lasiocarpos Cavanilles
Hibiscus moscheutos L.
Hibiscus laevis Allioni

BRASSICACEAE
Armoracia aquatica (Eaton) Wiegand
Nasturtium microphyllum (Boenninghausen) Reichenbach
Nasturtium officinale R. Braun

STYRACACEAE
Styrax americana Lam.

Rosidae

CRASSULACEAE
Tillaea aquatica L.

ROSACEAE
Rosa palustris Marshall
Crataegus aestivalis (Walter) Torr. & A. Gray

FABACEAE
Aeschynomene indica L.

PODOSTEMACEAE
Podostemon ceratophyllum Michx.

HALORAGACEAE
Myriophyllum aquaticum (Vell.) Verdc.
Myriophyllum heterophyllum Michx.
Myriophyllum laxum Shuttleworth ex Chapm.
Myriophyllum pinnatum (Walter) BSP
Myriophyllum spicatum L.

TABLE 5. continued

LYTHRACEAE
Decodon verticillatus (L.) Elliott
Lythrum lineare L.
Lythrum flagellare Shuttleworth ex Chapm.
Didiplis diandra (DC.) Wood

ONAGRACEAE
Ludwigia peruviana (L.) Hara
Ludwigia peploides (HBK) Raven
Ludwigia arcuata Walter
Ludwigia repens Forst.
Ludwigia palustris (L.) Elliott
Ludwigia uruguayensis (Camb.) Hara
Ludwigia alata Elliott
Ludwigia suffruticosa Walter

NYSSACEAE
Nyssa aquatica L.
Nyssa biflora Walter

AQUIFOLIACEAE
Ilex cassine L.
Ilex coriacea (Pursh) Chapm.

EUPHORBIACEAE
Caperonia castaneifolia (L.) St.-Hil.
Caperonia palustris (L.) St.-Hil.
Stillingia aquatica Chapm.

APIACEAE
Eryngium aquaticum L.
Lilaeopsis carolinensis Coult. & Rose
Lilaeopsis chinensis (L.) Kuntze
Hydrocotyle ranunculoides L. f.
Hydrocotyle umbellata L.
Ptilimnium fluviatile (Rose) Mathias
Ptilimnium nodosum (Rose) Mathias
Oxypolis greenmanii Mathias & Constance
Cicuta mexicana Coult. & Rose

Asteridae

MENYANTHACEAE
Nymphoides aquatica (Gmelin) Kuntze
Nymphoides cordata (Elliott) Fernald
Nymphoides peltata (Gmelin) Kuntze

OLEACEAE
Fraxinus caroliniana Miller

HYDROPHYLLACEAE
Hydrolea quadrivalvis Walter
Hydrolea uniflora Raf.

CALLITRICHACEAE
Callitriche heterophylla Pursh
Callitriche stagnalis Scop.

LAMIACEAE
Physostegia correllii (Lundell) Shinners

SCROPHULARIACEAE
Bacopa caroliniana (Walter) Robinson
Bacopa rotundifolia (Michx.) Wettst.
Amphianthus pusillus Torr.
Limnophila sessiliflora (Vahl) Blume
Limnophila indica (L.) Druce
Veronica analallis-aquatica L.

LENTIBULARIACEAE
Utricularia purpurea Walter
Utricularia radiata Small
Utricularia inflata Walter
Utricularia floridana Nash
Utricularia fibrosa Walter
Utricularia olivacea Wright ex Griseb.
Utricularia biflora Lam.
Utricularia foliosa L.

ACANTHACEAE
Hygrophila lacustris (Schlecht. & Cham.) Nees
Justicia americana (L.) Vahl

PLANTAGINACEAE
Plantago cordata Lam.

ASTERACEAE
Spilanthes americana (Mutis) Hieron.
Bidens laevis L.
Bidens cernua L.
Sclerolepis uniflora (Walt.) BSP

Liliiopsida

Alismatidae

POTAMOGETONACEAE
Potamogeton diversifolius Raf.
Potamogeton tennesseensis Fernald
Potamogeton epihydrus Raf.
Potamogeton confervoides Reichenbach
Potamogeton pusillus L.
Potamogeton foliosus Raf.
Potamogeton floridanus Small
Potamogeton crispus L.
Potamogeton amplifolius Tuckerman
Potamogeton pulcher Tuckerman
Potamogeton nodosus Poiret
Potamogeton gramineus L.
Potamogeton illinoensis Morong
Potamogeton perfoliatus L.
Potamogeton robbinsii Oakes
Potamogeton hillii Morong
Potamogeton oakesianus Robbins
Stuckenia pectinata (L.) Börner

HYDROCHARITACEAE
Najas marina L.
Najas guadalupensis (Sprengel) Magnus
Najas gracillima (Braun) Magnus
Najas filifolia Haynes
Najas minor Allioni
Najas wrightiana Braun
Hydrilla verticillata (L. f.) Royle
Elodea canadensis Michx.
Elodea nuttallii (Planchon) St. John
Limnobium spongia (Bosc) Richard ex Steudel
Egeria densa Planchon
Ottelia alismoides (L.) Persoon
Blyxa aubertii Richard
Vallisneria americana Michx.
Thalassia testudinum Banks ex König
Halophila johnsonii Eiseman
Halophila decipiens Ostenfeld
Halophila engelmannii Asch.

CYMODOCEACEAE
Halodule wrightii Asch.
Syringodium filiforme Kützing

ZOSTERACEAE
Zostera marina L.

LIMNOCHARITACEAE
Hydrocleys nymphoides (Willd.) Buchenau

ALISMATACEAE
Echinodorus tenellus (Martius) Buchenau
Echinodorus berteroi (Sprengel) Fassett
Echinodorus cordifolius (L.) Griseb.
Echinodorus floridanus Haynes & Burkhalter
Alisma subcordatum Raf.
Sagittaria subulata (L.) Bucheanu
Sagittaria guayanensis Kunth
Sagittaria montevidensis Cham. & Schltdl.
Sagittaria platyphylla (Engelm.) Smith
Sagittaria filiformis Smith
Sagittaria kurziana Glück
Sagittaria isoetiformis Smith
Sagittaria fasciculata Beal
Sagittaria secundifolia Kral
Sagittaria graminea Michx.
Sagittaria lancifolia L.
Sagittaria papillosa Buchenau
Sagittaria rigida Pursh
Sagittaria engelmanniana Smith
Sagittaria australis (Smith) Small
Sagittaria brevirostra Mackenzie & Bush
Sagittaria latifolia Willd.

RUPPIACEAE
Ruppia maritima L.

ZANNICHELLIACEAE
Zannichellia palustris L.

Arecidae

ARACEAE
Pistia stratiotes L.
Acorus calamus L.
Orontium aquaticum L.
Peltandra sagittifolia (Michx.) Morong
Peltandra virginica (L.) Schott & Endl.

LEMNACEAE
Spirodela polyrhiza (L.) Schleid.
Spirodela punctata (Meyer) Thompson
Lemna gibba L.
Lemna minor L.
Lemna perpusilla Torr.
Lemna trisulca L.
Lemna valdiviana Phil.
Wolffia columbiana Karst.
Wolffia papulifera Thompson
Wolffia punctata Griseb.
Wolffiella floridana (Smith) Thompson
Wolffiella gladiata (Hegelm.) Hegelm.
Wolffiella lingulata (Hegelm.) Hegelm.
Wolffiella oblonga (Phil.) Hegelm.

TABLE 5. continued

Commelinidae

POACEAE
Paspalum acuminatum Raddi.
Paspalum dissectum (L.) Linnaues
Paspalum repens Berg.
Zizaniopsis milacea (Michx.) Doell. & Asch.
Zizania aquatica L.
Hydrochloa caroliniensis Beauv.

CYPERACEAE
Dulichium arundinaceum (L.) Britton
Cyperus articulatus L.
Eleocharis interstincta (Vahl) R. & S.
Eleocharis equisetoides (Elliott) Torr.
Eleocharis quadrangulata (Michx.) R. & S.
Eleocharis robbinsii Oakes
Eleocharis elongata Chapm.
Websteria confervoides (Poiret) Hooper
Scirpus subterminalis Torr.
Scirpus etuberculatus (Steud.) Kuntze
Scirpus californicus (C. Meyer) Steud.
Scirpus validus Vahl
Scirpus cubensis Poepp. & Kunth
Scirpus lineatus Michx.
Rhynchospora traceyi Britton
Cladium jamaicense Crantz
Carex walteriana Bailey
Carex lurida Wahl.

TYPHACEAE
Typha latifolia L.
Typha angustifolia L.
Typha domingensis Persoon

SPARGANIACEAE (after Gleason & Cronquist 1991)
Sparganium americanum Nuttall
Sparganium eurycarpum Engelm.
Sparganium androcladum (Engelm.) Morong

MAYACACEAE
Mayaca fluviatilis Aublet

ERIOCAULONACEAE
Eriocaulon compressum Lam.
Eriocaulon cinereum R. Braun

JUNCACEAE
Juncus repens Michx.

Zingiberidae

MARANTACEAE
Thalia dealbata Roscoe
Thalia geniculata L.

Lilidae

PONTEDERIACEAE (after Rosatti 1987)
Eichhornia crassipes (Martius) Solms
Heteranthera dubia (Jacq.) MacM.
Heteranthera reniformis R. & P.
Heteranthera multiflora (Griseb.) Horn
Heteranthera rotundifolia (Kunth) Griseb.
Heteranthera limosa (Sw.) Willd.
Pontederia cordata L.

IRIDACEAE
Iris brevicaulis Raf.
Iris hexagona Walter
Iris virginica L.

ORCHIDACEAE
Platanthera repens (Nuttall) Wood.

REFERENCES

CRONQUIST, A. 1980. Asteraceae. Vol. I. In: A. E. Radford, et al., eds. Vascular flora of the Southeastern United States. Univ. North Carolina Press, Chapel Hill.

GLEASON, H.A. and A. CRONQUIST. 1991. Manual of vascular plants of northeastern United States and adjacent Canada, 2nd ed. New York Botanical Garden, New York.

GODFREY, R.K. and J.W. WOOTEN. 1979. Aquatic and wetland plants of southeastern United States. Monocotyledons. Univ. Georgia Press, Athens.

GODFREY, R.K. and J.W. WOOTEN. 1981. Aquatic and wetland plants of southeastern United States. Dicotyledons. Univ. Georgia Press, Athens.

HAYNES, R.R. 1968. *Potamogeton* in Louisiana. Proc. Louisiana Acad. Sci. 1:82–90.

HAYNES, R.R. 1998. Alabama. *Elatine triandra* Schkuhr (Elatinaceae). Castanea 63:81–82.

LLOYD, R.M. 1993. 12. Parkeriaceae Hooker. pp. 119–121. In N. Morin, et al., eds., Flora of North America, Vol. 2. Oxford Univ. Press, New York.

LUMPKIN, T.A. 1993. 25. Azollaceae Wettstein. In: N. Morin, et al., eds. Flora of North America, Vol. 2. Oxford Univ. Press, New York. Pp. 338–342.

NAUMAN, C.E. 1993. 24. Salviniaceae Reichenbach. In: N. Morin, et al., eds. Flora of North America, Vol. 2. Oxford Univ. Press, New York. Pp. 336–337.

Rosatti, T.J. 1987. The genera of Pontederiaceae in the southeastern United States. J. Arnold Arbor. 68:35–71.

TAYLOR, W.C., N.T. LUEBKE, D.M. BRITTON, R.J. HICKEY, and D.F. BRUNTON. 1993. 4. Isoëtaceae Reichenbach. In: N. Morin, et al., eds. Flora of North America, Vol. 2. Oxford Univ. Press, New York. Pp. 64–75.

THIERET, J.W., R.R. HAYNES, and D.H. DIKE. 1969. *Blyxa aubertii* (Hydrocharitaceae) in Louisiana: New to North America. Sida 3:343–344.

TUCKER, G.C. 1986. The genera of Elatinaceae in the southeastern United States. J. Arnold Arbor. 67:471–483.

MEDICINAL PLANTS OF THE SOUTHEASTERN UNITED STATES

EDWARD M. CROOM, JR.

National Center for the Development of Natural Products
School of Pharmacy
University of Mississippi
University, MS 38677, U.S.A.

ABSTRACT

Written records of the use of plants for improving health, and for preventing and treating disease, in the Southeastern United States began with the arrival of the first naturalists from Europe. In the 1800s, many plants were used in domestic practice and were popular as commercial and official therapeutic agents. In the first part of the 20th century, many plants from the Southeast were still popular medicines in both domestic and allopathic practice. However, by the last half of the 20th century, medicinal plants in the Southeast had been reduced to a very few commercial products and to a very limited use in domestic practice. The therapeutic uses of traditional medicines in the SE include the treatment of a large number of both acute and chronic conditions. These conditions are treated by a large diversity of native and introduced plants. Based on ethnobotanical research in the Southeast, the major medicinal plants still being used for self-treatment are discussed, as are the most important botanicals of commerce.

RESUMEN

Los registros escritos sobre el uso de las plantas para mejorar la salud, y para prevenir y tratar las enfermedades en el sureste de los Estados Unidos comenzó con la llegada de los primeros naturalistas de Europa. En el siglo XIX se usaron muchas plantas en la práctica doméstica y fueron populares como agentes comerciales y en la terapéutica oficial. En la primera parte del siglo XX muchas plantas del sureste eran todavía medicinas populares tanto en la práctica doméstica como en la alopática. Sin embargo, en la última mitad del siglo XX las plantas medicinales en el sureste han sido reducidas a unos pocos productos comerciales y a un uso muy limitado en la práctica doméstica. Los usos terapéuticos de las medicinas tradicionales en el SE incluyen el tratamiento de un gran número de enfermedades tanto agudas como crónicas. Estas enfermedades se tratan con una gran diversidad de plantas nativas e introducidas. Basados en la investigación etnobotánica realizada en el sureste, se discuten las principales plantas medicinales que todavía se usan en autotratamientos, ya que son los principales productos botánicos del comercio.

The flora of the Southeast includes the 14 states of Arkansas, Louisiana, Mississippi, Alabama, Georgia, Florida, North and South Carolina, Tennessee, Kentucky, Virginia, West Virginia, Maryland, and Delaware. The Biota of North America Program database reports the SE has: 246 families, 1,821 genera, and 7,341 species of plants (J. Kartesz, pers. comm., 3/27/97). Information on the plant medicines of the SE comes mainly from the following sources: European naturalists in the 18th century; from official, domestic & commercial remedies in the 19th century; as well as ethnographers & ethnobotanists in the 20th century. The European naturalists' accounts, such as John Lawson, would include reports of the medicinal and other useful plants they found in the exploration of the Southeast (Lawson 1967). In the 19th century, many of the popular medicinal plants used by physicians and pharmacists included natives of the Southeast (Wood & Bache 1839). Only in the late 19th and until the middle of the 20th century, did many of the professional reports by ethnographers and ethnobotanists document the major medicinal plants used by Native Americans in the Southeast. Major contributions for the Southeast included work on the Catawba, Cherokee, Choctaw, Creek, Houma, and Seminole (Mooney 1886; Mooney 1888; Mooney 1889; Mooney 1890,; Mooney; 1898; Mooney & Olbrechts 1932; Speck 1934; Speck 1937; Speck 1941; Speck 1944; Sturtevant 1955; Swanton 1918; Swanton 1928; Taylor 1940).

Many of the uses of plants reported from the 18th to the middle of the 20th century for the Southeast were reported as still being used in more recent works on medicinal plants by African Americans (Morton 1974), Appalachian whites (Bolyard 1981) and Native Americans (Croom 1982). Although many works have documented the medicinal and useful plants of Native Americans (for an overview of the entire North America see Moerman 1998), the medicinal uses of

SIDA, Bot. Misc. 18: 29–33. 2000

plants in the Southeast has been less thoroughly documented for African-Americans, whites, and other ethnic groups. Even during the past 25 years, it has been discovered that many rural residents of the Southeast such as the Lumbee Indians in the coastal plain of North Carolina have a knowledge of many medicinal plants (the Lumbee reported using over 80 species) that can be used to treat an extensive range of disorders as reported in Table 1 (adapted from Croom 1992). The diversity of uses of Southeastern medicinal plants used either in the recent past or popular today are illustrated by a number of examples in Table 2.

Plant medicines are made into teas, poultices, and salves. Most remedies are similar to drug specifics using a single species of plant. Other remedies follow a botanical polypharmacy approach with several different plant species included in the formulae. Unlike teas consumed as a daily beverage, most medicinal plant teas are not prepared as an infusion. In general, medicinal plant teas are gently boiled on a stove in a pan of water until the decoction is dark color or until half the water is lost, a process that may take from one to two hours. The decoctions are taken internally in doses from one teaspoon to one quart or more, although most doses are a swallow to a cup of liquid. The amount consumed may be reduced to one-half to three-fourths of the normal dose for those between ten and twenty years old. The tea is generally drunk hot, and the most important doses are considered the ones taken without food in the stomach before going to sleep at night and upon waking in the morning. When teas are not stored in the refrigerator, alcohol is added as a preservative.

Salves and ointments are prepared by boiling the plants in water until only a small amount of liquid remains. The decoction is then strained and the liquid is combined with hog or cow fat or a commercial product such as Petroleum Jelly.

A standard decoction is made by folding back the plant from the growing tip or end of a branch in six- to eight-inch lengths. These tightly folded bundles are measured for different diameters. The largest amount is measured by placing the index finger on the tip of the thumb; to make the next smallest size measurement, the index finger is placed on the first joint from the tip of the thumb (between the distal and proximal phalanx); the smallest hand measurement is made with the index finger placed at the base of the thumb (proximal phalanx).

The question addressing whether these traditional medicinal plants are safe and effective is well illustrated by the use of Poke (*Phytolacca americana* L.). Poke is a plant widely used in folk medicine as a spring "tonic" or blood purifier. Poke has a number of poisonous properties. In humans and animals Poke can cause burning of the mouth, cramps, vomiting blood, bloody diarrhea, visual disturbances, weakened respiration, and death (Kingsbury 1964; Hardin & Arena 1973). Handling the plant can cause a temporary rise in leukemia-like white blood cells (Barker et al. 1965). The known poisonous principles are triterpene saponins, protein mitogens, and oxalates. Even if Poke is effective in stimulating the immune system for fighting infections, can it be used safely? This depends on a number of factors.

The most important factors in Poke toxicity are which plant parts used, when it is harvested, how it is prepared, and the dosage. Young Poke leaves gathered in the spring and boiled in water which is then discarded, have been eaten by thousands of southerners as a food or a medicine without any symptoms of acute poisoning. The changes in leukocytes may be beneficial, acting, as many practitioners believe, as a spring tonic or blood purifier to rid the blood of infections. At least theoretically, the powerful mitogenic activity of Poke also may induce permanent leukemia or other types of cancer. These great potentials of Poke as rational therapeutic agent or insidious toxin are yet to be firmly established by studies that are based on the actual details of harvest, preparation, and administration as a food or domestic remedy.

Whether for personal or commercial use, the diversity of the plants used is better known than the extent of the use of southeastern plant species. In rural central Mississippi a random household in 1993 survey found that 71% of the respondents reported using at least one plant-derived remedy during the past year. The most popular plants used in the Mississippi survey during the previous year and their major uses were: 1) Lemon (47% of the respondents) juice internally for colds, cough, sore throat, and high blood pressure, 2) Aloe (27%) leaves externally for burns, 3) Castor Oil (14%) internally for constipation and colds; 4) Turpentine (12%) inter-

TABLE 1. Disorders treated by Lumbee Plant Medicines

Cancer
- External

Circulatory System
- Blood clots
- Dropsy
- Heart
- Hypertension

Digestive System
- Mouth
- Pyorrhea
- Sores
- Teeth cleaning
- Teething
- Thrush

Stomach/Intestine
- Colic
- Diarrhea
- Indigestion
- Laxative
- Parasites
- Ulcer

Endocrine, Metabolic, and Nutritional
- Disorders
- Diabetes
- Gout
- Pellagra

General Medicine
- Blood purifier
- Tonic

Infectious and Parasitic Diseases
- Chicken pox
- General infections
- Malaria
- Measles
- Scabies
- Whooping cough

Injury/Poisons/Toxins
- Antidotes
- Burns
- Emetics
- Skin irritation/abrasions
- Snake bites
- Spider bites
- Tick bites
- Wounds

Muscle, Skeletal, and Joint Disorders
- Arthritis/rheumatism
- Backache
- Spasms
- Sprains
- Swellings

Nervous System
- Analgesic
- Neuritis
- Sedative

Pharmaceutical Aid
- Flavoring

Reproductive System
- Aphrodisiac/sex tonic
- Birth control
- Female sexual organ & menstrual disorders
- Male sexual organ disorders
- Pregnancy/childbirth
- Venereal disease

Respiratory System
- Asthma
- Colds
- Cough
- Influenza
- Pneumonia

Skin
- Boils/sores
- Dandruff
- General & miscellaneous skin disorders
- Itching
- Jaundice
- Pimples
- Poison ivy/poison oak
- Rash
- Sunburn

Symptoms
- Fever
- Headache
- Pain

Urinary System
- Bladder
- Enuresis
- Kidney
- Kidney stones
- Urine, retention

Veterinary
- Intestinal parasites
- Laxative
- Snake bite
- Tonic

TABLE 2. Herbal medicines in the southeast United States.

Common name	Latin name	Main uses	Preparation
Calamus	*Acorus calamus*	Digestion, colds	Rhizome chewed or decoction
Catnip	*Nepeta cataria*	Make babies sleep	Sweetened leaf decoction
Chinaberry	*Melia azedarach*	De-worming Livestock	Fruits mashed and mixed with feed
Field Thistle	*Sonchus asper*	Teething	Roots as necklace
Gall Berry	*Ilex glabra*	Induce vomiting when poisoned	Berries swallowed
Jimsonweed	*Datura stramonium*	Asthma	Leaves smoked
John-the-Worker	*Hypericum hypericoides* & spp.	Urinary, skin, & digestive problems	Stem & leaf decoction
Mullein	*Verbascum thapsus*	Swellings, sprains	Leaf poultice
Oak	*Quercus* spp.	Sprains	Poultice of inner bark
Pine Top Tea	*Pinus* spp.	Colds & flu	Decoction of branch tips (stem & leaves)
Poke	*Phytolacca americana*	Blood purifier, spring tonic	Leaves boiled and eaten as greens
Purge Grass	*Sisyrichium* spp.	Laxative	Root decoction
Queen's Delight	*Stillingia sylvatica*	Venereal disease	Root decoction
Rabbit Tobacco	*Gnaphalium obtusifolium*	Colds, flu, cough	Brown leaves chewed or decoction
Rat's Vein	*Chimaphila maculata*	Kidney problems	Whole plant decoction
Sassafras	*Sassafras albidum*	Measles, colds, flu, spring tonic	Root bark decoction
Sumac, Salty Berry	*Rhus glabra* & *R. copallina*	Urinary, bed wetting	Berries sucked or cold water extract
Wicky	*Kalmia carolina*	Scabies	Stem & leaf decoction in bath water
Wild Horehound	*Eupatorium rotundifolium*	Colds, flu	Stem and leaf decoction
Yellow Root	*Xanthorhiza simplicissima*	Stomach & mouth ulcers	Rhizome chewed or cold water extract

nally for intestinal parasites, and externally for cuts and sores; 5) Tobacco (12%) leaves applied externally for bee and wasp stings, and 5) Garlic (10%) internally for high blood pressure (Frate et al. 1995).

Currently, southeastern plants used as Over-the-Counter drugs include Witch Hazel, *Hamamelis virginiana* L., as a topical astringent and Slippery Elm, *Ulmus rubra* Muhl., as a demulcent lozenge for soothing sore throats. Under current governmental law, more medicinal plants from the Southeast are sold as dietary supplements than as drugs. According to a recent estimate from the leading supplier of bulk botanicals in the Southeast, the world wide demand for the following Southeastern plants (in 1,000's of pounds) is as follows: American Ginseng, *Panax quinquefolius* L., 100–110; Blood Root, *Sanguinaria canadensis* L., 100–110; Black Cohosh, *Cimicifuga racemosa* (L.) Nutt., 250–300; Goldenseal, *Hydrastis canadensis* L., 300–350; Echinacea, *Echinacea angustifolia* DC., 200–250; Missouri Snake Root, *Parthenium integrifolium* L., 175–200; Passion Flower, *Passiflora incarnata* L., 450–500; Saw Palmetto, *Serenoa repens* (Bartr.) Small, 5,000; Slippery Elm, *Ulmus rubra* Muhl., 200–250; and Witch Hazel, *Hamamelis virginiana*L., 400–450 (Fletcher 1998).

SUMMARY

From the time of written records, our Southeast flora has been the source of a great diversity of medicinal plants for personal and commercial use. Although popular, the question of the actual safety and efficacy of medicinal plants has been a subject of continuous debate (Croom 1983). The continuation of the scientific study of these plants would seem promising for improving both the health and the economy of the Southeast and the world.

ACKNOWLEDGMENTS

Special thanks are given to John Kartesz, Chapel Hill, NC for information on the SE flora from the Biota of North America database. For information on the current world wide sales of plants native to the Southeast, special thanks to Ed Fletcher of Wilcox Drugs, Boone, NC.

INTERNET RESOURCES

Botanical Databases

USDA, NRCS 1997. The PLANTS database. (http://plants.usda.gov/). National Plant Data Center, Baton Rouge, LA 70874-4490. Can search by taxon and states, includes species distribution and maps.

Ethnobotany Database—Worldwide

Beckstrom-Sternberg, Stephen M., James A. Duke, and K.K. Wain. "The Ethnobotany Database." http://probe.nalusda.gov:8300/cgi-bin/browse/ethnobotdb. (ACEDB version 4.3 - data version July 1994). Search by taxon, use, country. Worldwide 80,000 records.

Native Americans—Ethnobotany Database

Daniel E. Moerman. "Native American Ethnobotany." foods, drugs, dyes, fibers and other uses of plants (a total of over 47,000 items). The database represents uses by 291 Native American groups of 3,895 species from 243 different plant families.

Native Americans—Medicinal Database

Beckstrom-Sternberg, Stephen M., Daniel E. Moerman, and James A. Duke. "The Medicinal Plants of Native America Database." http://probe.nalusda.gov:8300/cgi-bin/browse/mpnadb. (ACEDB version 4.3 - data version June 1995). Search by Author, tribe, use, or taxon.

Phytochemistry Database

Beckstrom-Sternberg, Stephen M., and James A. Duke. "The Phytochemical Database." http://probe.nalusda.gov:8300/cgi-bin/browse/phytochemdb. (ACEDB version 4.3 - data version July 1994). Search by activity, assay, plant chemical constituents, or taxon.

REFERENCES

BARKER, B.E., P. FARNES, and H. FANGER. 1965. Mitogenic activity in *Phytolacca americana*. Lancet 1:170.

BOLYARD J.C. 1981. Medicinal plants and home remedies of Appalachia. C.C. Thomas, Springfield, IL.

CROOM, JR., E.M. 1982. Medicinal plants of the Lumbee Indians. Dissertation. North Carolina State University, Raleigh.

CROOM, E.M., Jr. 1983. Documenting and evaluating herbal remedies. Econ. Bot. 37:13–27.

CROOM, E.M., Jr. 1992. Herbal medicine among the Lumbee Indians. In: Kirkland J, Mathews H.F., Sullivan C.W. III, Baldwin K., eds. Herbal and magical medicine. Duke University Press, Durham, NC. Pp. 137–69.

FLETCHER, E. 1998. Personal communication. Wilcox Drug, Boone, NC.

FRATE, D.A., E.M. CROOM, JR., J.B. FRATE, J.P. JUERGENS, and E.F. MEYDRECH. 1995. Self-treatment with herbal and other plant-derived remedies–rural Mississippi. Morbidity and Mortality Weekly Report 44(11):204–207, March 24, 1995.

HARDIN, J.W. and J.M. ARENA. 1973. Human poisoning from native and cultivated plants. 2nd ed. Duke University Press, Durham, NC.

KINGSBURY, J.M. 1964. Poisonous plants of the United States and Canada. Prentice-Hall, Englewood Cliffs, NJ.

LAWSON, J.A. New voyage to Carolina. 1967. University of North Carolina Press, Chapel Hill.

MOERMAN, D.E. 1998. Native American ethnobotany. Timber Press, Portland, OR.

MOONEY, J. 1886. The sacred formulas of the Cherokees. Bureau of American Ethnology. Smithsonian. 7th Annual Report. Washington, DC. Pp. 301–97.

__________. 1888. Myths of the Cherokees. J. Amer. Folklore 1:97–108.

__________. 1889. Cherokee plant lore. Amer. Anthropol. 2: 223–24.

__________. 1890. Cherokee theory and practice of medicine. J. Amer. Folklore 3:44–50.

__________. 1898. Myths of the Cherokee. Bureau of American Ethnology. Smithsonian Ann. Rep. 19.

MOONEY, J. and F.M. OLBRECHTS. 1932. The Swimmer Manuscript: Cherokee sacred formulas and medicinal prescriptions. Bureau of American Ethnology. Smithsonian Bull. 99. GPO.

MORTON, J.F. 1974. Folk remedies of the Low Country. Seeman, Miami, FL.

SPECK, F. G. 1934. Catawba texts. Columbia Univ. Contr. Anthropol. 24:1–91.

__________. 1937. Catawba medicines and vurative practices. Publ. Phil. Anthropol. Soc. 1:179–97.

__________. 1941. A list of plant curatives obtained from the Houma Indians of Louisiana. Primitive Man 14.4:49–73.

__________. 1944. Catawba herbals and vurative practices. J. Amer. Folklore 52:37–50.

STURTEVANT, W.C. 1955. The Mikasuki Seminole: Medical beliefs and practices. Dissertation. Yale University.

SWANTON, J.R. 1918. Catawba notes. J. Washington Acad. Sci. 8:623–29.

__________. 1928. Religious beliefs and medical practices of the Creek Indians. Bureau of American Ethnology. Smithsonian Ann. Rep. 42:472–672.

TAYLOR, L.A. 1940. Plants used as curatives by certain southeastern tribes. Botanical Museum of Harvard University, Cambridge, MA.

WOOD, G.B. and F. BACHE. 1839. The dispensatory of the United States of America. 4th ed. Grigg and Elliot, Philadelphia, PA.

A SURVEY OF THE HERBARIA OF THE SOUTHEAST UNITED STATES

V.A. FUNK and NANCY MORIN[1]

Department of Botany
Smithsonian Institution MRC 166
Washington D.C. 20560, U.S.A.
(funkv@nmnh.si.edu)

American Association of Botanical Gardens and Arboreta
351 Longwood Road
Kennett Square, PA 19348, U.S.A.

ABSTRACT

Herbaria are critically important resources for botanical research and teaching as well as for reference and comparison for those doing plant-related applied research and biological resource management. It is a matter of national importance that we maintain and support these facilities, especially at a regional level. This study surveyed the current status, activity level, and prognosis for future health of the herbaria in the southeastern region of the U.S. The results of the survey indicate that, of the 161 herbaria still active, more than 50% of the holdings are located in only seven herbaria. Eighteen percent of the herbaria are active or very active and 47% are moderately active. However, eight herbaria holding over 700,000 specimens have closed and the specimens either been discarded (6500) or transferred to other herbaria. In addition, 40 of the herbaria, one quarter of the active ones that we surveyed, holding nearly 1,600,000 specimens, are either inactive or the person in charge fears that they will become so in the near future. We encourage herbaria in the southeast U.S. to develop state networks and to actively work with administrators and politicians to encourage the support of herbaria, and we suggest that similar surveys be under taken in other regions in the U.S.

RESUMEN

Los herbarios son unas fuentes importantes para la investigación botánica y la enseñanza, así como para referencia y comparación para aquellos que hacen investigación aplicada relacionada con las plantas y manejo de recursos biológicos. Es un tema de importancia nacional que nosotros mantengamos y costeemos estas instalaciones especialmente a nivel regional. Este estudio analiza el estatus actual, nivel de actividad, y pronóstico de futuro de los herbarios de la región sureste de los Estados Unidos. Los resultados de este análisis indican que de los 161 herbarios todavía activos, más del 50% de los materiales están depositados en sólo siete herbarios. El dieciocho por ciento de los herbarios están activos o muy activos y el 47% están moderadamente activos. Sin embargo, ocho herbarios que contienen más de 700,000 especímenes han cerrado y los especímenes han sido eliminados (6500) o transferidos a otros herbarios. Además, 40 de los herbarios, un cuarto de los activos que han sido analizados, que contienen cerca de 1,600,000 especímenes, están inactivos o sus directores temen que lo estén en un futuro cercano. Animamos a los herbarios del sureste de Estados Unidos a desarrollar redes estatales y a trabajar activamente con la administración y los políticos para que mantengan los herbarios, y sugerimos que se realicen análisis similares en otras regiones de los Estados Unidos

INTRODUCTION

Herbaria are collections of preserved plant specimens. The specimens are usually pressed, dried, and mounted on stiff paper, but they are sometimes preserved in liquid or frozen. They are usually accompanied by collection data giving detailed location and habitat information, date collected, and more recently the latitude and longitude. Some plant collections are hundreds of years old and others are being collected today. Herbarium specimens provide the comparative material that is essential for correctly identifying and classifying plants, but because of the time span and the associated data, these collections also are a remarkable, irreplaceable source of information about the plants and the conditions that exist, and have existed, in the world.

Herbarium specimens can be used for many purposes, a few are listed below:

1. The most common use of herbarium specimens is to discover or confirm the identity of an unknown plant. Someone from the state extension service, or the local horticulture club, or the

[1]Present address: The Arboretum at Flagstaff, 4001 S. Woody Mountain Road, Flagstaff, AZ 86001, U.S.A.

SIDA, Bot. Misc. 18: 35–52. 2000

native plant society, or the forest service, or the biology department, finds a plant and, out of curiosity or a need to manage that plant or the landscape it grows in, wants to find out what it is. Finding out what it is generally takes for granted that a huge amount of work has already been done: that the area where the plant grows has been explored and collected, that the resulting specimens have been studied and classified by a specialist, that a monograph or flora has been written in which the plant is named, described, and identifiable in keys. But for the average plant finder, the important thing is that once the plant has been provisionally identified, it is possible to go to a herbarium, compare it with the several candidate species, and determine by that comparison which species it is.

2. Specimens document which plants grew where through time. This information makes it possible to deduce what environmental conditions once existed in a particular area, including such things as climate and the presence of heavy metals, and it is possible to document what conditions a plant species required.

3. Specimens document what plants grew with what other plants. From these data, it is possible to determine what the vegetation was like, what the structure of the plant communities was like, and sometimes what animals were also likely to have been present.

4. Specimens document the morphology and anatomy of individuals of a particular species from which one can determine whether the morphology has changed through time or if it varies with different growing conditions.

5. Specimens provide temporal and spatial information. Because each collection usually has not only the location but also the date it was collected and because they are usually collected in reproductive stages, they are very useful for investigating such topics as changes in plant distribution through time, phenology, and biogeography.

6. Specimens document the concepts of the taxonomists and other biologists who have studied and commented on the specimens. This function is perhaps best illustrated by type specimens, because types are the final arbiter of the application of a scientific name. But it is also possible to read a paper by a taxonomist long dead and examine the same specimens that he or she examined and annotated in order to better understand taxonomic concepts and use of terminology.

7. From some specimens one can extract and analyze the DNA and/or analyze the chemical composition in order to understand taxonomic relationships, prove accumulation of metals or other substances, or screen for pharmaceutical effectiveness.

8. Specimens of the same species collected through the years provide the historical and current distribution of a plant and one can chart the changes in its distribution. This might elucidate the pattern of spread of a noxious weed or the encroachment on territory of a plant of conservation concern, but even in less dramatic cases it helps us understand how the world is changing.

9. Microscopic observations of herbarium specimens can provide information for many interesting studies of hairs, pollen, stomata, vasculature, and a wide range of other plant attributes.

Herbaria with large national and international collections are important for answering questions and solving problems on a global scale. However, it is important to have local and regional herbaria, not only because a local herbarium is more likely to have plants from the local area, but also someone who needs to identify a plant is unlikely to want take the plant to an out-of-area herbarium. Students learning their local flora need a herbarium close by for comparison and study, as well. In addition, any training for careers in environmental studies from preparing environmental impact studies to working with fish and wildlife projects requires a knowledge of the local flora and the ability to identify local plants.

There are more than 60 million specimens in 628 herbaria in the U.S., and 7 million specimens in 110 herbaria in Canada, according to Index Herbariorum (Holmgren et al. 1990). Over nine and one half million herbarium specimens are held in herbaria in the southeastern U.S., and significant collections of southeastern plants are also found in major herbaria elsewhere, particularly Missouri Botanical Garden, The New York Botanical Garden, and the U.S. National Herbarium in the National Museum of Natural History. These are a very important national and international resource, critically needed for the study and conservation of the plants and animals of the southeastern U.S.

The purpose of this study was to conduct a survey of the herbaria of the southeast U.S. and to produce a document that would examine the activity of those herbaria and provide information to the botanical community on the availability of specimens and information. In the end we also gained useful information on the status of taxonomy and some insight into the number of herbaria that either need or are likely to need a new home.

We defined the southeast to include 13 states: Alabama, Arkansas, Florida, Georgia, Kentucky, Louisiana, Mississippi, North Carolina, South Carolina, Tennessee, Texas, Virginia, and West Virginia. We included West Virginia because it is located between Kentucky and Virginia. Rather than try to separate east Texas, we included all of Texas, in part because many participants in the BRIT sponsored symposium were from Texas but mainly because it was too difficult to draw boundaries regarding holdings.

Herbaria have a long history in the southeast U.S. Fourteen of the 183 southeastern herbaria were established before 1900, with the oldest, established in 1771, being in Salem College in Winston-Salem (SC) and the second in 1820 at the Charleston Museum (CHARL). Between 1900 and 1950 another 50 herbaria were opened but the real peak of growth in numbers of herbaria was between 1950 and 1975, when more than 80 herbaria were established. The vast majority of these are associated with universities or colleges, and their history and use is associated with teaching and local floristic projects. Many of the herbaria have rich local collections not duplicated elsewhere. Although some herbaria have been transferred to other homes, and some have completely disappeared, many are actively growing, loaning specimens, and undertaking research projects. Except for the largest herbaria, most are curated by faculty or staff who also have teaching responsibilities, and some are cared for entirely by volunteers, in some cases faculty from other departments. This paper discusses the geographical coverage of various herbaria and their current research projects, as well as future prospects for herbaria in the southeast.

There are several herbaria that have large southeastern holdings that were not included in this survey. The U.S. National Herbarium at the National Museum of Natural History, Smithsonian Institution, Washington, D.C. (U.S.) has extensive holdings from the southeast U.S. but they make up a small part of the total number of specimens (ca. 5 million). Statistics in this paper are for total holdings, and including U.S. would skew the analyses significantly. However, taxonomists working on the flora of the southeastern U.S. should be aware that there is a separate herbarium at US for plants of the District of Columbia and vicinity that consists of approximately 56,000 specimens. In addition, the absence of the herbaria of eastern Oklahoma and the herbarium of Southern Illinois University in Carbondale has been pointed out by several individuals from Oklahoma and Illinois. All of these herbaria have good holdings of southeastern plants and should be consulted when studying plants from the southeast.

Our goals in this project were the following:

1. Assess the activity and well-being of herbaria in the southeast U.S.
2. Make observations on the findings and possibly make helpful suggestions,
3. Provide a document that would be an accurate evaluation and summary,
4. Provide an electronic web site with information that would increase loans and other activity of southeast U.S. herbaria,
5. Inspire others to conduct similar surveys in other regions of the U.S., and
6. Call attention to the importance of herbaria and provide documentation that can be used when herbarium supporters are meeting with administrators.

METHODS

A questionnaire was developed in September 1997 and sent to all 183 southeast herbaria listed in the on-line version of Index Herbariorum (Holmgren et al. 1998). There are many small herbaria that are not listed in the Index Herbariorum, indeed several who attended this symposium asked about becoming a part of the Index Herbariorum, but we had no way of knowing about these herbaria and so they were not included. Any herbarium interested in being listed in the Index Herbariorum should contact Patricia Holmgren at the New York Botanical Garden (Bronx, NY 10458, pholmgren@nybg.org). It is hoped that this paper will encourage all of the unregistered herbaria to become members of this Index.

We recognized that everyone is busy and kept the survey questions to a minimum. The following five questions were included in a letter (samples of all questionnaires can be found on the web site):

1. What important collections are held in your herbarium (e.g., what regions, what taxonomic groups, and/or which collectors)?
2. How "active" is your herbarium? Are there incoming and outgoing loans? Are new specimens being added to the collections? Are the collections being actively curated?
3. Who is in charge of the herbarium (please give us name and title)? Are they full-time, part-time, or spare-time?
4. What active research and/or collecting programs are currently underway in the herbarium (if any)?
5. What do you think the future of this herbarium is–steady or rapid growth? maintaining the status quo? likely to be needing a new source of support or institutional home?

The response to the initial letter, sent to the attention of the curator of the herbarium, was about 50%. A second letter, sent to the herbaria that had not responded to the first letter was addressed to the Herbarium without specifying a person, provided responses from an additional 20% of the total. The remaining approximately 45 herbaria were reached by telephone. Some curators were easy to reach and just very busy, others were more difficult to find and were located through Department Heads or Deans. Many of the herbaria where it was difficult to obtain information were either no longer open or had been dispersed / discarded. Ultimately information was gathered for all herbaria.

As the responses were analyzed it became apparent that certain essential information had not been requested, such as the current number of collections in the herbarium and a specific question on whether or not the curator had any release time from teaching to take care of the herbarium. Some people included this information, but for others we asked by telephone or used information from Index Herbariorum or colleagues. Because the questionnaire had asked for the number of specimens added each year, we were able to adjust the IH number to reflect the approximate number of additions that would be in the collection today. The estimates were conservative and as a result the figures for some herbaria may be underestimated. After much discussion an "optimal questionnaire" has been developed (Table 1) for use as a starting point in similar surveys elsewhere.

Many curators provided detailed information as well as publications and website addresses. In addition, many curators were candid about the current status and future of their herbaria. As information accumulated the institutions were evaluated and given an informal activity code. There were seven "levels of activity," from those that had been destroyed (-1) to those that were very active in terms of acquisitions, loans, research projects, numbers of students and number of faculty (5). The criteria for each activity code value are listed in Table 2. These categories are not based on size but rather on the amount of use the collections receive and how well they are housed and curated. The survey forms and the activity codes assigned were reviewed by several people who wish to remain anonymous. In addition, a draft of this manuscript was sent to 20 of the institutions that were surveyed, at least one in every state. The authors considered all comments, but take full responsibility for the results presented in the paper.

All data from the survey were entered into an Access© database. Missing information was filled in from Index Herbariorum or by personal communications from individuals who knew the institution. This full database is available on the web site of the first author (http://mason.gmu.edu/~ckelloff/vfunk/) and eventually it will be listed on BRIT website (www.brit.org).

TABLE 1: Optimal questionnaire recommended for future herbarium surveys.

(if information is from IH please indicate with *) Acronym:
Who filled out this form ______________________________ State:
Activity Code:
# of collections

COLLECTIONS:
What percentage of the collection do you consider to be well curated?
Specimens:
- Total number of collections in the herbarium:
- Approximate percentage from state:
- Approximate percentage from region (incl. State):
- Approximate percentage from outside of USA:
- Is there a cryptogamic or fungi collection, if so what taxa and what size:

People (special collections):

Areas
- states:
- geographic region:

Taxa
- taxonomic groups:

Activity:
- # of Loans per year (number or qualitative answer):
- # of exchanges per year (number or qualitative answer):
- # of specimens added per year (number or qualitative answer):

Additional Information:

CURATION (a curator is someone who touches herbarium specimens lovingly):
Faculty Curator (name; title; birthdate; Full-time curator, part-time, spare- time, retired) (does job as curator give you a reduced teaching load?)
Staff Curator (name; birthdate; full-time; part time; spare-time)
Other faculty or staff or associates that are interested and support the activity of the herbarium.
Do you have student workers (y/n):
Who provided the information if different from above.

USES:
Research (y/n) and what projects:
Teaching (y/n) and how many courses involve the herbarium:
Agency Use (USDA; FWS; National Park Service; state agencies; etc.):
Voucher collections (y/n) and what projects:
Training of Undergraduates (y/n) and how many on average:
Training of Graduate students (y/n) what level and how many on average:
Have you or are you computerizing any of your collections (y/n) and if so how many have been done and what are plans for future?

SUPPORT:
What are your main sources of support (University, State, Endowment, Private donations, etc.)?
Do you receive outside funding for the herbarium (grants or contracts), for specific research projects?

FACILITIES:
Is your space and storage capacity adequate for the next 10 years?

COLLABORATION AND ASSOCIATION:
Universities (do not include your home institution):
Botanical Gardens (do include one that is at your home institution):
Museums:
Agencies of Federal or State Government:
Garden or herb Clubs:

FUTURE:
What is the future of the herbarium: rapid growth, moderate growth, status quo, low activity or no activity, needs a new home or is likely to need a new home in the next 5 years.
Please answer the following question it is important: "What will happen to the herbarium when the curator retires or leaves?"
ADDITIONAL COMMENTS:

TABLE 2: Criteria for Activity Codes used for Herbaria.

-1. Dispersed to unknown locality, known to have been thrown away, or missing (it is not there and current faculty have no idea where it is).
0. Transferred

Table 2 continued

1. Essentially closed; no activity; not used; no students working on projects or theses; no faculty curator, no staff; no growth; loosing ground; needs a new home or may need a new home in the near future.
2. Open; little or hardly any activity; maybe low level of use as reference or teaching collection; no projects or theses; maybe a spare time faculty or retired faculty; no staff; little or no growth; barely holding on to current level or sliding a bit; future of collection uncertain or may need a new home in the future.
3. Open; some moderate activity; used for some teaching and research; several theses and projects; a spare-time or part-time faculty curator and possibly a part-time or full-time staff person; gradual growth or status quo; collection secure for now; space OK or somewhat tight.
4. Open; active; used for teaching and research; theses and projects; at least a part-time faculty curator and a staff person; yearly increase of 1-several thousand collections; moving forward; collection secure for now; space OK. [possible for this to be a "one man band"]
5. Open; very active; research facility, may also be used for teaching; lots of projects and theses; faculty and staff; yearly increase of several thousand collections; leaping ahead; collections secure; institutional commitment for adequate space and support. [must have institutional support and more than one faculty interested in the herbarium, not a "one man band"]

Appendix A is a subset of the data matrix that includes the name of the institution, state, abbreviation, number of specimens, and activity code.

RESULTS AND DISCUSSION

Of the original 183 herbaria, 161 are now known to be extant (Table 3). Sixteen were transferred and six disappeared. The 161 herbaria house a total of 9,646,505 specimens (probably underestimated and does not include backlog). We recognize that some of these herbaria, most notably TEX, BRIT, DUKE, and IBE, have substantial holdings that are not from the southeastern U.S., however the majority of the specimens are from this area.

In the United States there are approximately 60,000,000 plant collections housed in 628 herbaria. About 20 million are in large herbaria that have substantial collections from outside of the U.S. which leaves 40 million specimens from the U.S. The United States covers 3,028,825 sq. miles (minus Alaska). If one divides 40 million by the sq. miles of the U.S. (minus Alaska) the result is 13.21 plant collections per square mile. There are 9,646,505 plant collections in the southeast U.S. and 813,486 square miles of territory, or 11.85 plant collections per square mile, which is not so far from the national average. Our impression is that coverage of Europe in European herbaria is higher.

The 161 herbaria in southeast U.S. are found predominately in public and private universities and colleges with approximately 105 in state supported schools and approximately 37 in private ones. Five herbaria are in free-standing Museums and eight are under federal or state government control. Three herbaria are housed in freestanding botanical gardens: Fairchild Tropical Garden, Selby Botanical Gardens, and Birmingham Botanical Gardens.

The largest herbaria in the southeast U.S. (in thousands of specimens) are TEX 1,100, BRIT 860, DUKE 700, and NCU 630. These top four herbaria have approximately one third of the specimens housed in the southeast U.S. They are followed in size by TENN 470, FLAS 400, and NLU 350. The top seven herbaria have approximately one half of the collections housed in the southeast U.S. Other herbaria that have over 100,000 specimens include IBE (253), USF (223), TAES (200), LSU (185), FSU (183), GA (180), WVA (160), LTU (122), NCSC (120), NO (108), VPI (105), LAF (104) and UARK (100). Therefore, the top 18 herbaria have a total of 6,550,000 specimens, over two thirds of the collections held in the southeast U.S.

The Activity Level can be examined in several ways. Figure 1 illustrates the number of herbaria by activity level. It represents a bell-shaped distribution in which 47% of the herbaria have activity level 3 while 26% are in category 2, 9% in category 1 and 18% are operating at a levels 4 or 5. Examining the number of specimens housed in herbaria with each activity code shows a very different picture with a linear relationship (Fig. 2) where 68% of the specimens are housed in herbaria operating at levels 4 or 5, and 10% are found in herbaria operating at levels 1 or 2. In other words, while most of the herbaria have activity level 3, most of the specimens are in institutions with level 4 or 5. Still, 9% of the total number of specimens is around 900 thousand.

TABLE 3: Total number of herbaria and the number of herbaria/number of thousands of specimens listed by state and activity code. Numbers of specimens were rounded to nearest thousand, 500's were rounded up.

State	No. Herb.	5	4	3	2	1
Alabama	8	-	1/50	4/74	2/10	1/65
Arkansas	8	-	1/16	5/111	1/5	1/100
Florida	17	4/753	1/30	8/244	2/9	2/16
Georgia	11	-	1/180	6/114	2/30	2/10
Kentucky	8	-	1/40	7/198	-	-
Louisiana	13	-	2/535	7/303	3/147	1/15
Mississippi	6	-	3/334	1/19	1/11	1/3
North Carolina	12	1/700	3/698	4/173	3/20	1/?
South Carolina	8	-	1/55	4/119	2/45	1/1
Tennessee	8	1/470	-	5/55	2/40	-
Texas	30	2/1,960	2/225	12/393	12/184	2/8
Virginia	23	-	3/203	7/194	10/93	3/36
West Virginia	8	-	1/160	5/98	2/17	-
Total	160	8/ 3,883	20/ 2,526	75/ 2,095	42/ 611	15/ 254

If one examines the data by state certain trends appear. In Table 3, which gives the total number of collections by state and by activity level, it is easy to see that some states have many more specimens than others. For instance, Texas has the most herbaria (30) and the most specimens (3,019,000). Florida has the largest number of herbaria (4) that have an activity code of 5. West Virginia is the smallest state but South Carolina has the fewest number of herbarium specimens. Most states have only a few herbaria in categories 1 or 2, except for Texas with 14 out of 30 (47 %) and Virginia with 13 out of 23 (61%). Table 4 gives the number of specimens per state and per square mile and shows that North Carolina is the best collected state with 33.15 collections per square mile. The least well-collected states are Alabama and Arkansas, both with around 4 specimens per square mile. Table 5 combines the number of herbaria, number of specimens and number of specimens per square mile to show that only North Carolina, Louisiana, and Florida have numbers that are significantly greater than the national average and six states are significantly below that average.

Evaluation by State

ALABAMA.—Most of the herbaria in Alabama are moderately active. Two have been transferred from their original institution within state. The herbaria seem to have focused largely on collections within the state, with northern, northwestern, central, and southeast Alabama being covered. Two herbaria have general southeast U.S. collections and there is one horticultural herbarium. The largest herbarium in Alabama, Auburn University (AUA) with 65,000 collections, is currently inactive, although a newly hired faculty member, a molecular systematist, may change this. The University of Alabama herbarium (UNA), however, has new, state of the art collections facilities, and it and the Jacksonville State University herbarium (JSU) have active graduate programs.

ARKANSAS.—In Arkansas, the herbarium at Henderson University (HEND) is very active, and the state is fortunate to have five additional moderately active herbaria. Unfortunately, the herbarium at the University of Arkansas, Fayetteville (UARK), the largest, with 100,000 specimens, is not active and its future is uncertain. The curator has retired, the specimens are being put into storage during a renovation project, and there is no guarantee that the herbarium will be restored to an active condition. There are two herbaria in the state with bryophyte collections and three doing natural area surveys (see note on page 52 for recent information).

FLORIDA.—Florida is the state with the greatest floristic diversity in the southeast and it has many herbaria. Five of the Florida herbaria are hardly active, eight are moderately active, one is very active, and four are extremely active. Three of the inactive herbaria focus on algae, and there should be some concern for their future. The University of Florida (FLAS) is the largest herbarium in the state with 400,000 specimens, and FLAS and Selby Botanical Gardens (SEL)

TABLE 4: Number of herbaria, number of specimens housed in each state, and number of specimens per square mile of area within that state, ordered by number of specimens.

STATE	NO. of HERB.	NO. of SPECIMENS	SQ. MILES in STATE	SPECIMENS/ SQ. MILE
TEXAS	30	3,019,000	267,339	11.29
NORTH CAROLINA	12	1,626,955	49,067	33.15
FLORIDA	17	1,049,250	58,560	17.92
LOUISIANA	13	999,200	48,523	20.59
TENNESSEE	8	565,300	42,244	13.38
VIRGINIA	23	520,820	40,817	12.76
MISSISSIPPI	6	367,100	47,716	7.69
GEORGIA	11	333,300	58,876	5.66
WEST VIRGINIA	8	275,000	24,181	11.37
KENTUCKY	8	237,500	40,395	5.88
ARKANSAS	8	231,500	53,104	4.36
ALABAMA	8	198,600	51,609	3.85
SOUTH CAROLINA	8	220,000	31,055	7.08

TABLE 5: Number of specimens per square mile of area within each state, ordered by state.

	Specimens per Sq. Mile		Specimens per Sq. Mile
North Carolina	33.15	Mississippi	7.69
Louisiana	20.59	South Carolina	7.08
Florida	17.92	Kentucky	5.88
Tennessee	13.38	Georgia	5.66
Virginia	12.76	Arkansas	4.36
West Virginia	11.37	Alabama	3.85
Texas	10.29		

have strong support for their herbaria. FLAS has a range of special collections including bryophytes, lichens, fungi, wood, and paleobotanical specimens. It should be noted that most of the major herbaria in Florida tend to include a research and acquisition focus outside of the U.S. The curators at the herbaria of Florida Atlantic University (FAU) and Florida State University (FSU), both of which have been expanded by very active taxonomists, have indicated that the long-term future should be of concern. In addition, the Myxomycete collection at the University of Central Florida (FTU) may need a new home.

GEORGIA.—Georgia has six moderately active herbaria and four that are hardly active. The University of Georgia (GA) is the largest, with 180,000 specimens, and it is also the most active. The University of Georgia (GAM) has an Ascomycete collection of potential importance that may need more support. Also of concern is the herbarium at Valdosta State University (VSC), which has nearly 50,000 specimens, but little institutional support.

KENTUCKY.—Kentucky has good state coverage, with seven moderately active herbaria. The University of Kentucky (KY) and Eastern Kentucky University (EKY) are the largest, with 40,000 specimens; Eastern Kentucky University seems to be the most active. The curator at Berea College (BEREA) will not be replaced when he retires, and the herbarium is likely to need a new home. All but one of the herbaria in Kentucky are essentially "one man bands."

LOUISIANA.—In Louisiana, there are seven moderately active, 4 hardly active, and 2 quite active herbaria. The largest is Northeast Louisiana University (NLU) with 350,000 specimens. Louisiana has herbaria with special collections in lichens and bryophytes; Louisiana State University (LSUS) has a Basidiomycete and Myxomycete collection. The Mycological herbarium at Tulane University (NO) is likely to need a new home. The University of Southwestern Louisiana (USLH), with 104,000 specimens and an important bryophyte collection, has an uncertain future.

MISSISSIPI.—Mississippi has three hardly active, one moderately active, and three quite

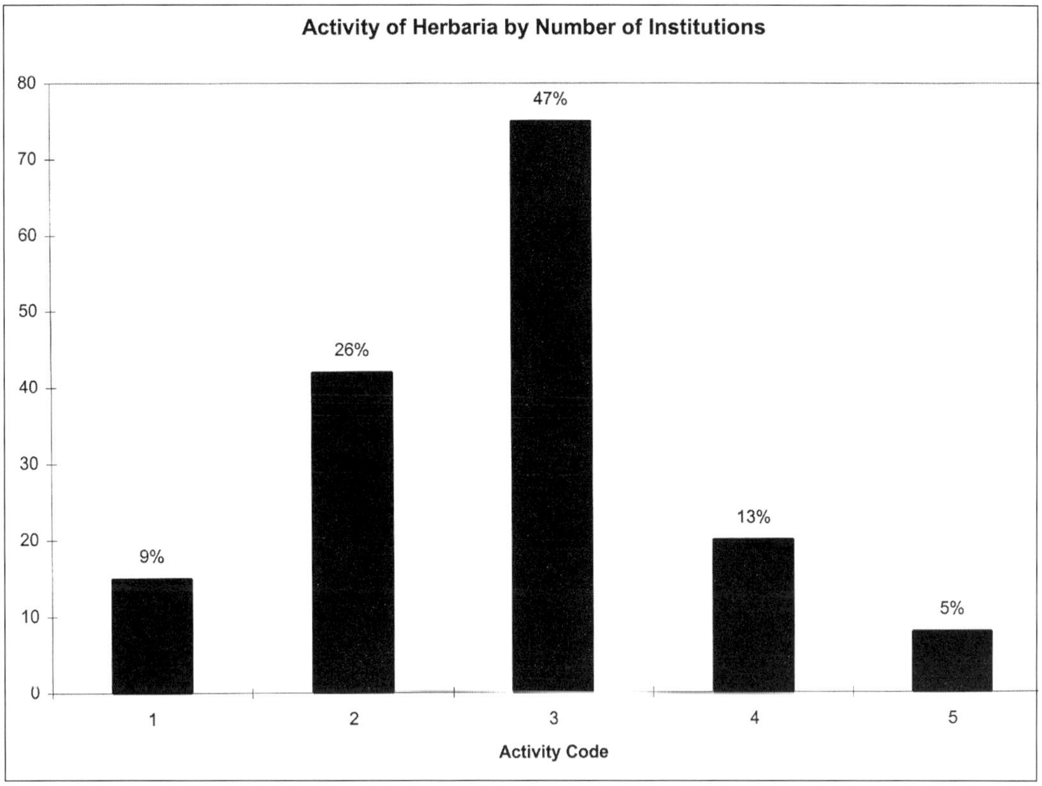

FIG. 1: The number of herbaria at each activity level.

active herbaria. The largest and most active herbarium, the Institute for Botanical Exploration (IBE), has a large South America collection. The Southern Weed Science Laboratory (SWSL), in Stoneville Mississippi has a special collection of sedges (Cyperaceae). The Gulf Coast Research Laboratory in Ocean Springs (HGCRL) has collections from the coastal plain.

NORTH CAROLINA.—North Carolina has had three of its herbaria transferred, two within the state and one to NY and five of its herbaria are hardly active. However, there are three moderately active, three active, and one very active herbarium as well. Duke University (DUKE) is largest, with 700,000 specimens, and is the most active. With the exception of DUKE, which has a large number of collections from Costa Rica, the herbaria in North Carolina focus on local and regional flora, with several specializing in collections from National Parks. There are lichenological and bryological collections in the state with DUKE having excellent cryptogamic and fungi collections. There was some indication that the Pfieffer University Herbarium (PFC) might need a new home but some funds have recently been allocated (pers. comm. R. Wilbur). Also, recently Appalachian State University (BOON) has hired two new taxonomists, one for vascular plants and one for fungi, and the systematic program is rapidly improving.

SOUTH CAROLINA.—South Carolina has three hardly active, four moderately active, and one quite active herbaria. The largest is University of South Carolina (USCH), with 80,000 specimens followed by Clemson (CLEM) with 55,000. The Charleston Museum (CHARL) has very early collections, although these are not available on loan. There are bryophyte, fungi, and algae collections in SC.

TENNESSEE.—The University of Tennessee (TENN) is the largest herbarium in the state, with 470,000 specimens. It is very active and is undergoing rapid growth, and it recently received an NSF facilities improvement grant. There are three hardly active and five moderately active herbaria in the state; several of them focus on park and forest inventories. The second largest herbarium in the state (VDB) was recently transferred to BRIT because the host department had little interest in maintaining organismal biology at the University.

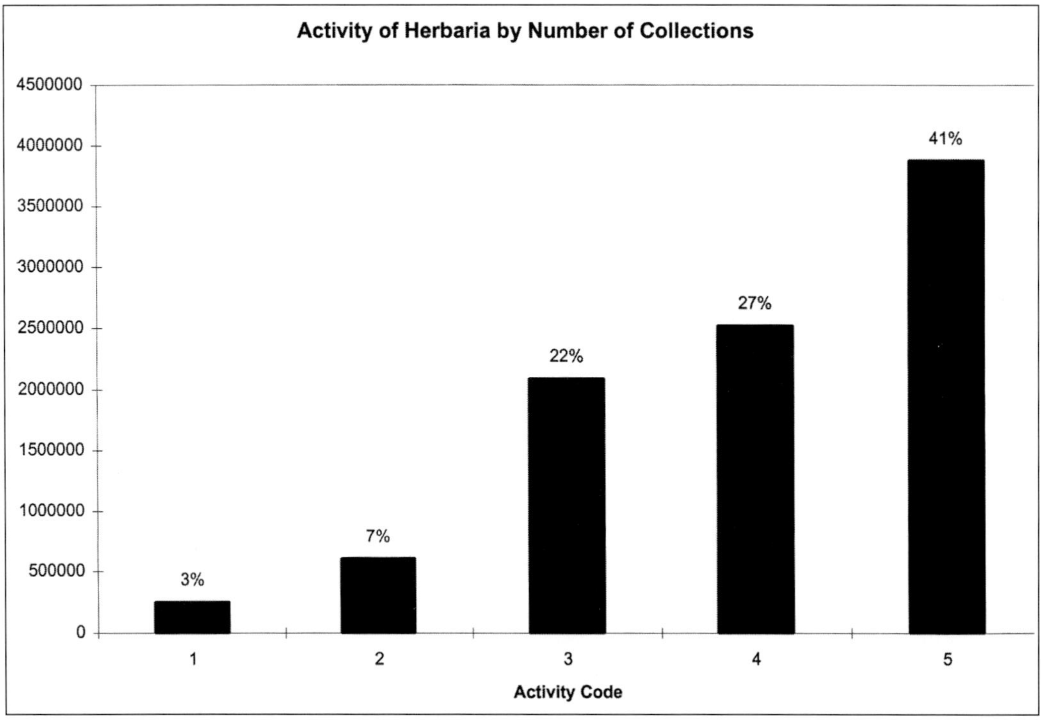

FIG. 2: The number of specimens found in the herbaria at each activity level.

TEXAS.—Texas covers a vast area and many habitats. The flora of central and west Texas is not strictly of the southeastern U.S., and certainly the largest herbaria found at the University of Texas (TEX) and the Botanical Research Institute of Texas (BRIT) have extensive collections from outside the state, however for the purposes of this paper it is really impossible to separate out those collections. Six herbaria have been transferred, all within the state. But in addition, there are 14 herbaria that are barely active and one might expect that many of these will need a new home soon. There are 12 herbaria that are moderately active, two that are quite active, and two that are very active. Early collections at Our Lady of the Lake University (LLC) and at the Fort Worth Museum of Science and History (FWM) may be of interest. Texas has bryological, lichenological, macroalgae, and rust fungi collections. The herbarium at South Plains College (SPLT) seems to be in danger and may be transferred elsewhere.

VIRGINIA.—The herbarium at Virginia Polytechnic Institute (VPI) is the largest with 98,000 specimens. There are 7 moderately and 3 quite active herbaria. The focus of most of the collections is on local floras and state parks, but there are also special horticultural, forestry, Myxomycete, and bluegreen algae collections. Virginia has an alarming figure of 13 herbaria, with a total of 129 thousand specimens, that are hardly active and may eventually need new homes. Certainly the future of the herbarium at the Virginia Military Institute (VMIL with 30,000 specimens) and the herbarium at the Virginia Department of Agriculture (VDAC with 34,000 specimens) seems to be uncertain.

WEST VIRGINIA.—The largest herbarium in West Virginia is the University of West Virginia, Morgantown (WVA) with 160,000 specimens, which has an active program with a lot of community interaction. In addition, there are two hardly active and five moderately active herbaria. Fairmont State College (FWVA) has a Myxomycete collection.

Only 20 herbaria mentioned specifically having cryptogamic collections. Ten of these have bryological collections, four have lichenological collections. Harbor Branch Oceanological Institute (HBFH) has an algal collection, and University of North Carolina at Moorhead (IMS) has collections of marine fungi and macro fungi. The herbaria at the Western Kentucky University

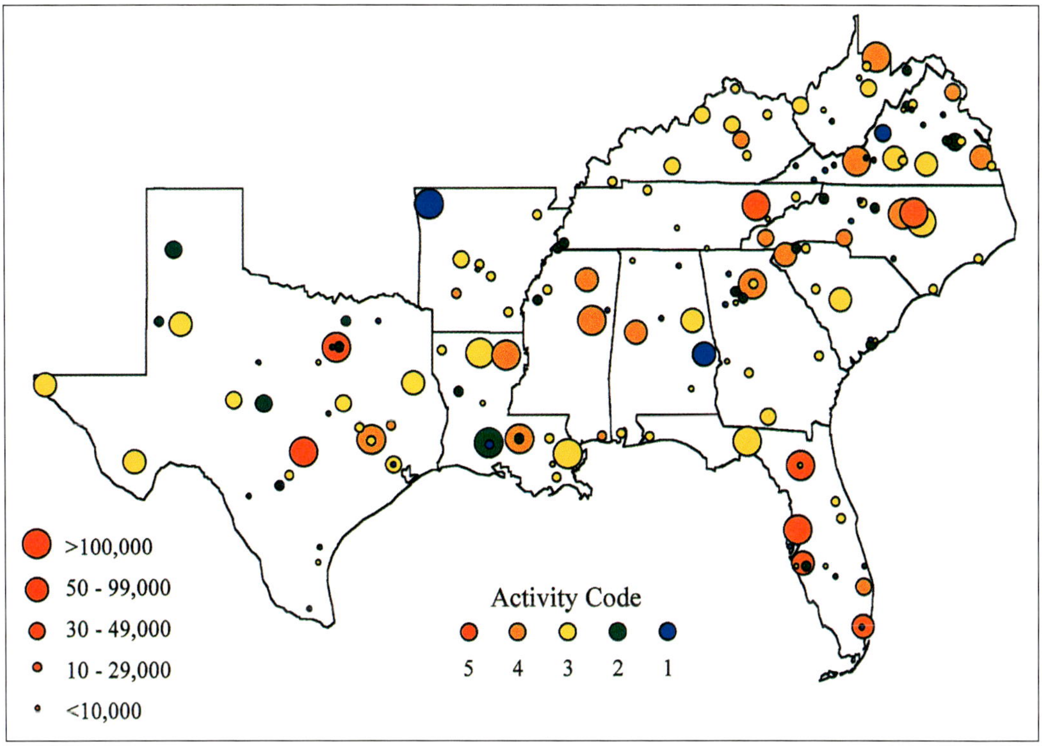

FIG. 3: A map of southeast US showing the location of the herbaria surveyed in this study. The size of the circle indicates the number of specimens and the color of the circle indicates the activity level.

(WKU) and the University of Richmond (URV) have Myxomycete collections, while the one at the University of Georgia (UGM) has an Ascomycete collection and Louisiana State University herbarium (LSUS) has a Basidiomycete collection. DUKE University has a large and diverse cryptogamic and fungi collection and graduate training program.

In general the small and medium sized herbaria that stay active do so by using students and volunteers. They have at least one (usually only one) faculty member that generates all of the interest. They produce local floras, checklists, state park surveys and other means to keep the herbarium active. Most larger herbaria are at least at a level 3 with AUA and UARK being the exceptions. AUA has just hired a molecular systematist with a taxonomic background so there is some hope that this collection will once again be active but current reports on UARK are not encouraging (see note on page 52 for recent information).

A map showing locations of herbaria in the Southeast U.S. is shown in Figure 3. The size of the circles reflects the number of specimens housed in the collection and the color indicates the activity code of that herbarium. The map shows that there are certain areas that have few or no active herbaria—in particular central and western Kentucky and Tennessee, southern Mississippi and Alabama, Arkansas, coastal South Carolina, southern West Virginia, eastern Kentucky, and west central Texas.

There are approximately 40 herbaria for which concern was expressed in the "Future" section of the form (Table 1). For these herbaria, the most frequent comment in this section was that the curator did not know what would happen to the herbarium when he/she retired. Because we did not directly ask this question in the first letter, it is likely that it is more common than the survey indicates. This question has been added to the form in Table 1 for all future surveys. However, if we look just at the 40 that expressed concern, they represent 25% of the herbaria that are still in existence in the southeast U.S. and hold approximately 1.6 million specimens. Small herbaria, and cryptogamic and fungi collections seem to be especially vulnerable. How-

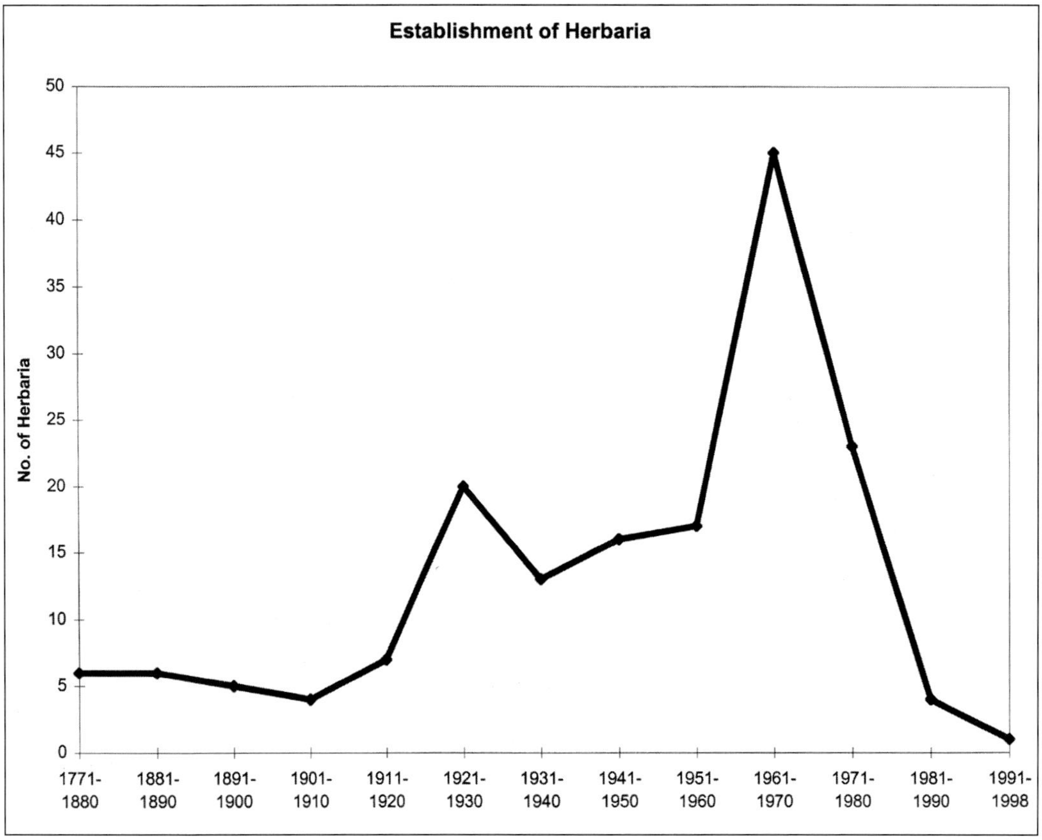

FIG. 4: The number of herbaria established in ten year increments.

ever, there are larger herbaria in this group as well, including four of the 17 with over 100,000 specimens: FSU, NLU, NO, and LAF.

By examining in 10 year intervals the number of herbaria founded (Fig. 4) it is clear that changes are taking place in the way herbaria are valued. The period between W.W.I and the great depression shows a surge, the period of the late sixties and early seventies when universities and science programs were expanding shows remarkable growth, but then a drop to nearly zero takes place, with just one new herbarium being founded since 1990 and that one is private (BRCH). The closing of herbaria reported in Index Herbariorum and this survey, and the predicted closing based on comments made in this survey (Fig. 5) show a troublesome future. There is the potential that at least 40 herbaria might close over the next few years. This is a far higher rate than any time in the past. In addition, if one examines Figure 6 it is clear that if all 40 herbaria close, the botanical community will have to find a home for at least 1,600,000 specimens, not including the backlog of unprocessed material. When one examines these trends it seems clear that unless some action is taken in the next few years, a large number of herbaria will close, requiring that their specimens be moved elsewhere and depriving students and communities of access to a valuable resource. These tables and graphs may indicate a broader problem—the general decline in plant taxonomy in Universities and Colleges throughout the southeast and as such it needs further investigation.

CONCLUDING REMARKS

It is hoped that this document will provide a practical guide to the collections, projects, and expertise held in the herbaria in the southeastern U.S. By consulting the database, researchers can see which herbaria have depth in their holdings of specific regions or taxa, and this should

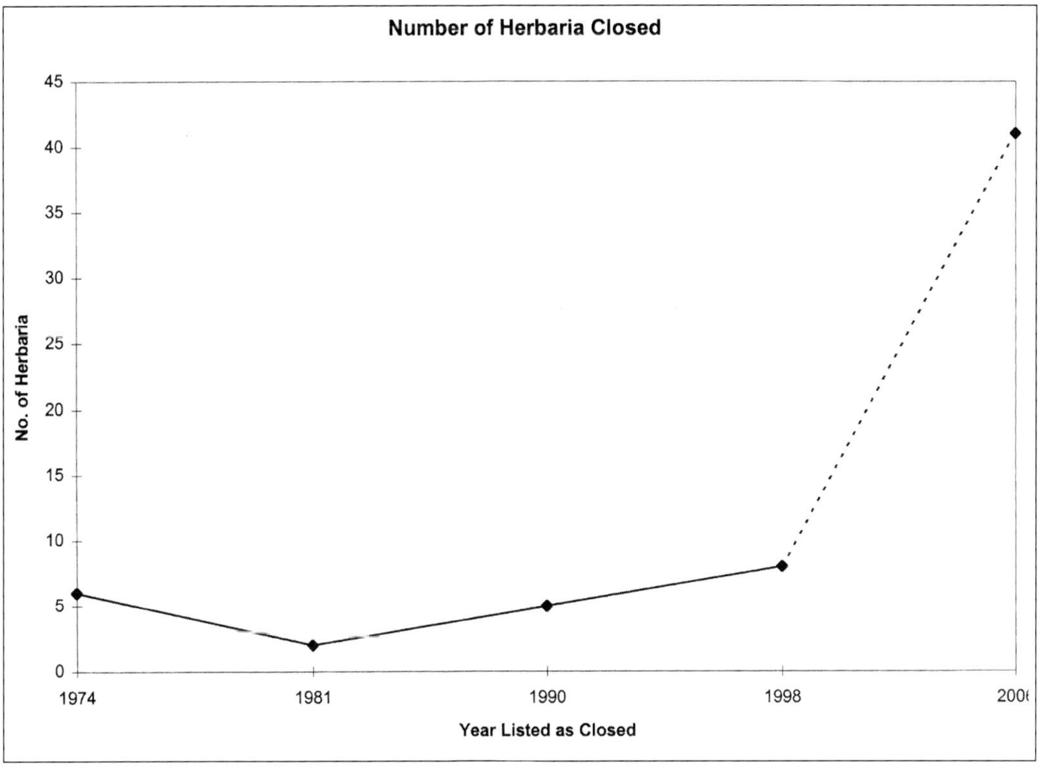

FIG. 5: Herbaria that have closed (solid line) and projected closing of herbaria (dashed line); based on information from Lanjouw & Stafleu 1964, Holmgren & Keuken 1974, Holmgren et al. 1981, Holmgren et al. 1990, Holmgren et al. 1998, and Funk & Morin, 1998 this survey.

guide them in borrowing specimens for study. The database may also facilitate development of collaborative projects.

The Convention on Biological Diversity, which endorsed a Global Taxonomy Initiative to promote capacity building in taxonomy, calls for each country to have a clearing-house mechanism to coordinate access to and distribution of taxonomic information in systematic collections. Within the U.S., Systematics Agenda 2000 confirms the importance of collections and taxonomic expertise to our understanding of biodiversity. The collections and expertise are essential for the training of taxonomists for the future as well. An international meeting of systematists in Darwin, Australia, recently declared the critical importance of evaluating the current status of collections and of specialists and of planning for their future health and growth. Systematists have an obligation to provide the information necessary for understanding global environmental processes and maintaining a national interest in documenting and studying our native flora. The results reported here clearly show that although there is much to be optimistic about, there is also reason to be concerned about the future prospects for herbaria and the capacity to do botanical research, at least in the southeast U.S.

It is hoped that this document will be useful for those who need to make compelling arguments to their deans, chairs, or funding bodies about the true local, regional, national, and international value of their collections, and the current assessment of their status. Many herbaria have been supported in part by federal or state funds and data such as these should prove useful at that level. Although in most cases ownership of a particular collection is clear, it is important for administrators to know that these collections have regional, national, and global importance and there should be a sense of community responsibility for them. In some cases they provide the only documentation for the historic makeup of the flora of an area. In other cases they may

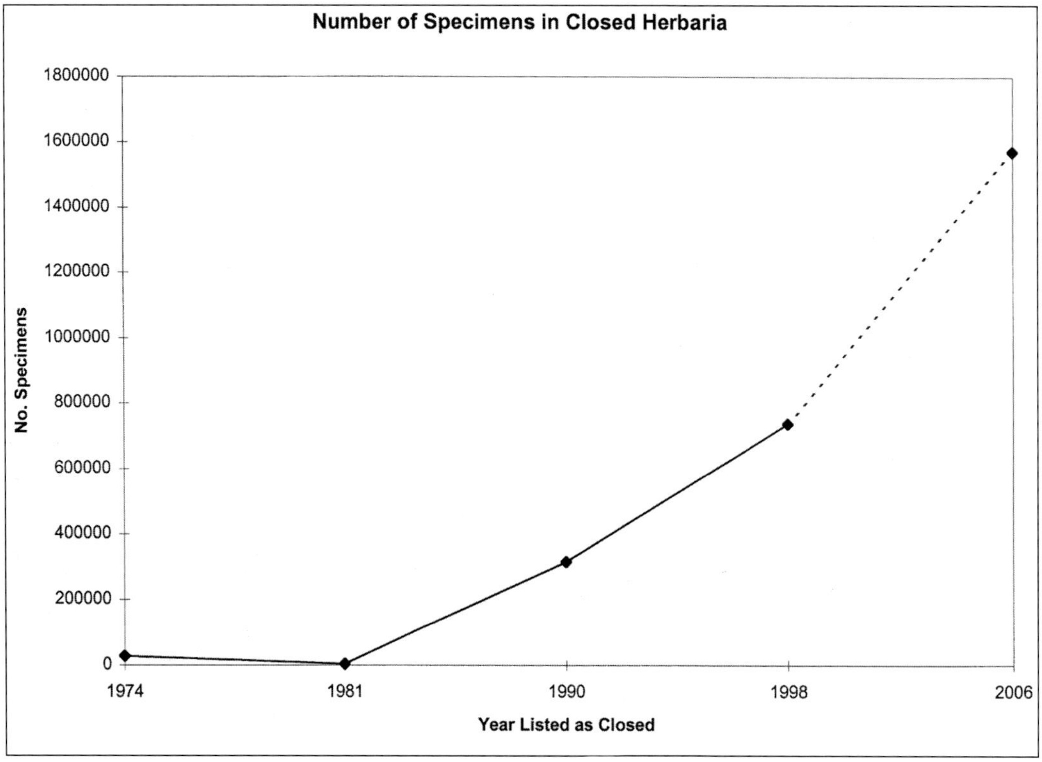

FIG. 6: The number of specimens found in herbaria that have closed (solid line) and herbaria that are projected to close (dashed line); based on information from Lanjouw & Stafleu 1964, Holmgren & Keuken 1974, Holmgren et al. 1981, Holmgren et al. 1990, Holmgren et al. 1998, and Funk & Morin, 1998, this survey.

house the only extant specimens of a rare or endangered species. Because of their regional and national importance, we think that a mechanism should be established to alert the systematics community in the event that a collection is endangered or about to be dispersed or discarded.

It would be useful for taxonomists in all states to form local networks or support groups so they can monitor the status of the herbaria within their state and so that they can call on colleagues to help convince officials at a College or University of the importance of their herbarium. It is necessary to stress the importance of local and regional collections. It is only by scientists joining forces with other herbarium users that one can hope to convince administrators that herbaria are important. In addition, it is important to convey a positive attitude about the work of the herbarium, regardless of its size or its activity level. Students and volunteers can have a great sense of pride in working with a local herbarium, and this pride can be fostered through newsletters, field trips, etc. Very small institutions can still have very useful and impressive websites. Many colleges and universities have botanical gardens or arboreta that benefit from associated herbaria. These sites are important in student training and the importance of the herbarium can be featured in public outreach programs. Garden clubs, local parks, local nature clubs and other groups can all be active partners in supporting and fighting for the herbarium. Because so many herbaria have been formed as part of a university or college, we tend to think of them entirely as academic; in fact many organizations, including departments of conservation, parks departments, non-governmental conservation organizations, and others, have a real need for a local herbarium, and sometimes they can be a source of funding as well.

We realize that one of the greatest challenges faced by institutions maintaining herbaria is funding. Because of this need, this paper will be provided to funding agencies to give them a snapshot of the current status of collections so they can see the magnitude of the problem of coverage and regional representation.

We think there would be real value in surveying the herbaria (and other kinds of systematics collections) in all regions of the U.S. in order to get a better understanding of our national capacity to document, study, and understand the organisms of our country. With this in mind, the authors are available to assist any individuals who wish to start such a survey. At least one other similar survey has been completed. It covers southern Africa (Smith & Willis 1997) and although it is a completely different area of the world it will be useful for comparison purposes.

APPENDIX A

Appendix A: Subset of data matrix containing the University or Institution name, state in which it is located, official abbreviation, number of specimens found in the herbarium, and the activity code (see Table 2). Complete data matrix can be found on the website (http://mason.gmu.edu/~ckelloff/vfunk). Numbers of specimens in parentheses are not included in totals because they have disappeared or been transferred.

University	State	Herb	Size	Code
Alabama Museum of Natural History	AL	ALU	(2,500)	0
Auburn University	AL	AUA	65,000	1
Birmingham Botanical Gardens	AL	BBG	5,100	2
University of Alabama, Huntsville	AL	HALA	5,000	2
Jacksonville State University	AL	JSU	50,000	3
Saint Bernard Abbey	AL	SB	(4,500)	0
Troy State University	AL	TROY	2,500	3
University of Alabama, Tuscaloosa	AL	UNA	50,000	4
University of North Alabama	AL	UNAF	7,000	3
University of South Alabama	AL	USAM	14,000	3
Arkansas Tech University	AR	APCR	35,000	3
Henderson State University	AR	HEND	15,500	4
Hendrix College	AR	HXC	5,000	2
University of Arkansas at Little Rock	AR	LRU	16,000	3
Arkansas State University	AR	STAR	25,000	3
University of Arkansas at Monticello	AR	UAM	21,000	3
University of Arkansas, Fayetteville	AR	UARK	100,000	1
University of Central Arkansas	AR	UCAC	14,000	3
Archbold Biological Station	FL	ARCH	6,500	3
University of Miami	FL	BUS	12,000	1
Stetson University	FL	DLF	10,500	3
Florida Atlantic University	FL	FAU	30,000	4
University of Florida	FL	FLAS	400,000	5
Florida Southern College	FL	FSCL	7,500	3
Florida State University	FL	FSU	183,000	3
Fairchild Tropical Garden	FL	FTG	64,000	5
University of Central Florida	FL	FTU	10,000	3
Gesneriad Research Foundation	FL	GES	4,300	3
Harbor Branch Oceanographic Institution	FL	HBFH	6,100	2
Mote Marine Laboratory	FL	MOT	2,600	2
Division of Plant Industry	FL	PIHG	7,500	3
Selby Botanical Gardens	FL	SEL	65,000	5
University of South Florida, St. Petersburg	FL	SPMS	3,750	1
Florida Marine Research Institute	FL	STPE	(1,000)	0
University of South Florida, Tampa	FL	USF	223,500	5
University of West Florida	FL	UWFP	13,000	3
Columbus State University	GA	COLG	5,000	3
Agnes Scott College	GA	DECA	2,500	3
USDA Forest Service	GA	FSSR	11,000	2
University of Georgia, Botany	GA	GA	180,000	4
Georgia Agricultural Experiment Station	GA	GAES	-	0
University of Georgia, Plant Pathology	GA	GAM	25,000	3
Georgia Southern University	GA	GAS	20,000	3
Emory University	GA	GEO	18,800	2
Georgia Southwestern State University	GA	GSW	13,000	3

University	State	Herb	Size	Code
Shorter College	GA	SHOR	4,000	1
Valdosta State University	GA	VSC	48,000	3
State University of West Georgia	GA	WGC	6,000	1
Berea College	KY	BEREA	22,000	3
University of Louisville	KY	DHL	35,000	3
Eastern Kentucky University	KY	EKY	40,000	4
Centre College of Kentucky	KY	KBRYO		-1
Northern Kentucky University	KY	KNK	24,500	3
University of Kentucky	KY	KY	40,000	3
Morehead State University	KY	MDKY	10,000	3
Murray State University	KY	MUR	28,000	3
Western Kentucky University	KY	WKU	38,000	3
University of Southwestern Louisiana, Biology	LA	LAF	104,000	2
Louisiana State University, Baton Rouge, Plants	LA	LSU	185,100	4
Louisiana State University, Baton Rouge, Mycology	LA	LSUM	23,000	2
Louisiana State University, Shreveport	LA	LSUS	20,000	3
Louisiana Tech University	LA	LTU	122,600	3
Northwestern State University	LA	NATC	20,000	2
Northeast Louisiana University	LA	NLU	350,000	4
Tulane University	LA	NO	108,000	3
University of New Orleans	LA	NOLS	9,500	3
Southeastern Louisiana University	LA	SELU	12,000	3
Southern Forest Experiment Station	LA	SFRP	5,000	3
Nicholls State University	LA	THIB	25,000	3
University of Southwestern Louisiana, Horticulture	LA	USLH	15,000	1
Delta State University	MS	DSC	11,000	2
Gulf Coast Research Laboratory	MS	HGCRL	20,000	4
Institute for Botanical Exploration	MS	IBE	253,500	4
University of Mississippi	MS	MISS	61,000	4
Mississippi State University	MS	MISSA	(53,000)	0
Mississippi University for Women	MS	MSCW	2,600	1
Southern Weed Science Laboratory	MS	SWSL	19,000	3
Duke University	NC	ABSM	(15,000)	0
Davidson College	NC	DCH	(16,155)	0
Duke University	NC	DUKE	700,000	5
University of North Carolina, Morehead City	NC	IMS	19,000	3
North Carolina A & T State University	NC	NCATG	10,000	2
North Carolina State University	NC	NCSC	120,000	3
University of North Carolina, Chapel Hill	NC	NCU	630,000	4
Pfeiffer University	NC	PFC	?	1
St. Andrews Presbyterian College	NC	SAPCL	9,000	2
Salem College	NC	SC	1,000	2
Southeastern Forest Experiment Station	NC	SEFES	-	0
University of North Carolina, Charlotte	NC	UNCC	38,000	4
Western Carolina University	NC	WCUH	30,000	4
Wake Forest University	NC	WFU	16,000	3
University of North Carolina, Wilmington	NC	WNC	17,800	3
Charleston Museum	SC	CHARL	25,000	2
Southern Research Station	SC	CHAS	1,000	1
The Citadel	SC	CITA	7,000	3
Clemson University	SC	CLEMS	55,000	4
Furman University	SC	FUGR	20,000	2
Newberry College	SC	NBYC	12,000	3
University of South Carolina	SC	UCSC	20,000	3
University of South Carolina	SC	USCH	80,000	3
Austin Peay State University	TN	APSC	23,000	3
East Tennessee State University	TN	ETSU	12,000	3

University	State	Herb	Size	Code
Great Smoky Mountain National Park	TN	GSMNP	7,300	3
Rhodes College	TN	MEM	20,000	2
Middle Tennessee State University	TN	MTSU	6,000	3
University of Memphis	TN	SWMT	20,000	2
University of Tennessee, Knoxville	TN	TENN	470,000	5
University of Tennessee, Chattanooga	TN	UCHT	7,000	3
Vanderbilt University	TN	VDB	(300,000)	0
Stephen F. Austin State University	TX	ASTC	75,000	3
Baylor University	TX	BAYLU	30,000	3
Botanical Research Center	TX	BRCH	15,000	3
Botanical Research Institute of Texas	TX	BRIT	860,000	5
Texas A&M University, Commerce	TX	ETST	2,500	1
Texas Christian University	TX	FW	500	2
Fort Worth Museum of Science and History	TX	FWM	12,000	2
University of Mary Hardin-Baylor	TX	HABAYC	2,200	2
Howard Payne University	TX	HPC	40,000	2
Houston Museum of Natural Science	TX	HPM	5,000	1
Hardin-Simmons University	TX	HSU	3,000	2
University of Texas, Austin	TX	LL	(65,000)	0
Our Lady of the Lake University	TX	LLC	10,000	2
University of North Texas	TX	NTSC	16,000	2
University of Texas-Pan American	TX	PAUH	8,000	2
Runyon Herbarium	TX	RUNYON	-	0
Angelo State University	TX	SAT	30,000	3
Vines Environmental Science Center	TX	SBSC	30,000	3
Sam Houston State University	TX	SHST	25,000	4
Southern Methodist University	TX	SMU	(410,000)	0
South Plains College	TX	SPLT	20,000	2
Science Research Center	TX	SRCG	-	0
Sul Ross State University	TX	SRSC	60,000	3
Southwest Texas State University	TX	SWT	10,000	3
Tarleton State University	TX	TAC	1,300	3
Texas A&M University, College Station, Range Science	TX	TAES	200,000	4
Texas A&M University, Kingsville	TX	TAIC	7,000	3
Texas A&M University, College Station, Biology	TX	TAMU	29,000	3
Texas Women's University	TX	TCSW	-	-1
University of Texas, Austin	TX	TEX/LL	1,100,000	5
Blackland Experiment Station	TX	TEXA	(5,300)	0
Prairie View A & M University	TX	TPV	-	-1
Texas Tech University	TX	TTC	50,000	3
U. S. Soil Conservation Service	TX	USSC	-	-1
University of Texas, Richardson	TX	UTD	(250,000)	0
University of Texas, El Paso	TX	UTEP	56,000	3
Southwest Texas Junior College	TX	UVST	6,000	2
West Texas A&M University	TX	WTS	60,000	2
Welder Wildlife Foundation	TX	WWF	5,600	2
Bridgewater College	VA	BDWR	4,300	3
Clinch Valley College	VA	CVCW	4,000	2
Emory and Henry College	VA	EHCV	2,500	1
Longwood College	VA	FARM	75,000	3
Rappahannock Community College South Campus	VA	GLEN	(1,100)	-1
George Mason University	VA	GMUF	35,000	4
Eastern Mennonite University	VA	HAVI	4,000	2
James Madison University	VA	JMUH	16,000	2
Lord Fairfax Community College	VA	LFCC	15,000	3
Lynchburg College	VA	LYN	51,500	3
Mary Washington College	VA	MWCF	4,000	2
Old Dominion University	VA	ODU	20,000	3

University	State	Herb	Size	Code
Randolph-Macon Woman's College	VA	RMWC	13,000	3
Virginia Western Community College	VA	ROAN	1,200	2
Radford University	VA	RUHV	5,500	2
Roanoke College	VA	SARC	3,500	2
Sweet Briar College	VA	SWBR	(3,620)	0
University of Richmond	VA	URV	15,000	3
University of Virginia, Mountain Lake	VA	VA	3,000	2
Virginia Commonwealth University	VA	VCU	17,000	2
Virginia Department of Agriculture	VA	VDAC	34,000	2
Virginia Military Institute	VA	VMIL	30,000	1
Virginia Polytechnic Institute	VA	VPI	105,000	4
Virginia State University	VA	VSUH	(4,300)	-1
College of William and Mary	VA	WILLI	62800	4
Wytheville Community College	VA	WYCO	2,500	1
Fairmont State College	WV	FWVA	12,000	3
Marshall University	WV	MUHW	45,000	3
University of Charleston	WV	MVC	5,000	3
Potomac State College	WV	PSP	15,000	2
Salem College	WV	SAWV	6,000	3
West Virginia University, Morgantown	WV	WVA	160,000	4
West Virginia Institute of Technology	WV	WVIT	2,000	2
West Virginia Wesleyan College	WV	WVW	30,000	3

Postscript.—The following information about UARK appeared after review and revision of the present manuscript. The duties and responsibilities for collections management and curation of UARK have been transferred to the University Museum. A curator has been appointed who has floristic and monographic experience as well as herbarium management experience. The Museum's Curator of Collections and Registrar has assumed part of the collections management activities. The herbarium will move to another location on campus in June 2000.

ACKNOWLEDGMENTS

We appreciate the help of G. McKee and J. Pruski (US) in mailing the letters and helping to telephone the herbaria. Also, thanks to T. Hollowell for help with the graphics. Our sincere thanks to all of the faculty, curators, and staff at the herbaria who have helped us in compiling this information. The manuscript was sent to over 20 individuals, at least one in every state in the southeast U.S. We thank the following eight individuals for their thoughtful comments: T. Bradley (GMUF), D. Ford-Werntz (WVA), B. Henson (USF), Z. Murrell (BOON), J. Townsend (CLEM), R. Wilbur (DUKE), and M. Woods (TROY). The original set of all of the completed forms has been offered to the Archives at the New York Botanical Garden, the home of Index Herbariorum.

REFERENCES

LANJOUW, J. and F.A. STAFLEU. 1964. Index Herbariorum, Part I: The herbaria of the world, 5th ed. Utrecht. [Regnum Veg. vol. 31]

HOLMGREN, P.K. and W. KEUKEN. 1974. Index Herbariorum, Part I: The herbaria of the world, 6th ed. Utrecht. [Regnum Veg. vol. 92]

HOLMGREN, P.K., W. KEUKEN, and E.K. SCHOFIELD. 1981. Index Herbariorum, Part I: The herbaria of the world, 7th ed. Utrecht. [Regnum Veg. vol. 106]

HOLMGREN, P.K., N.H. HOLMGREN, and L.C. BARNETT. 1990. Index Herbariorum, Part I: The herbaria of the world, 8th ed. Bronx. [Regnum Veg. vol. 120]

HOLMGREN, P.K., N.H. HOLMGREN, and L.C. BARNETT. 1998. On line updated version of Index Herbariorum, Part I: The herbaria of the world, 8th ed. [http://www.nybg.org/bsci/ih/ih.html

SMITH, G.F. and C.K. WILLIS. 1997. Index herbariorum: southern African supplement. Southern African Botanical Diversity Network, Report No. 2. SABONET; Pretoria, South Africa.

THE PLANTS DATABASE: A NATIONAL TAXONOMIC STANDARD

MARK W. SKINNER and J. SCOTT PETERSON

USDA–Natural Resources Conservation Service
National Plant Data Center
PO BOX 74490
Baton Rouge, LA 70874, U.S.A.

ABSTRACT

PLANTS (USDA, NRCS 1999) provides nomenclature and distribution for all terrestrial plants and lichens for the United States, and currently has data on more than 36,000 taxa. PLANTS is a World Wide Web (WWW)-based system reached at *http://plants.usda.gov*, and it is maintained by the USDA-Natural Resources Conservation Service (NRCS) National Plant Data Center. The PLANTS checklist serves as a national taxonomic standard for plant names, and is widely used by other federal agencies to standardize plant information and ease data sharing. The information in PLANTS—especially distribution, growth habit, vernacular name, and wetland, noxious weed, and endangerment status—makes it a valuable tool for those engaged in floristic and monographic studies of southeastern US plants. Major improvements are planned for 1999 and 2000, including significant data updates, improved appearance and reports, and more sophisticated download routines.

KEYWORDS: PLANTS, plant checklists, vascular plants, lichens, hornworts, liverworts, hepatics, mosses, North America, United States, digital, USDA, NRCS

RESUMEN

PLANTS (USDA, NRCS 1999) proporciona la nomenclatura y distribución de todas las plantas terrestres y líquenes de los Estados Unidos, y normalmente tiene datos de más de 36,000 taxa. PLANTS es un sistema basado en la World Wide Web (WWW) que puede encontrarse en *http://plants.usda.gov*, y que es mantenido por el USDA- Servicio de Conservación de Recursos Naturales (NRCS) del Centro de Datos Nacional de Plantas. El catálogo PLANTS sirve como un estándar taxonómico nacional para los nombres de las plantas, y es usado ampliamente por otros organismos federales para estandarizar la información sobre las plantas y facilitar el uso conjunto de datos. La información en PLANTS—especialmente distribución, forma de crecimiento, nombre popular, si es mala hierba de humedales, y grado de peligro—hace de él un instrumento valioso para aquellos que se involucran en estudios florísticos y monografías de las plantas del sureste de los Estados Unidos. Las mayores mejoras están planeadas para 1999 y 2000, incluyendo la puesta al día de datos importantes, mejora del aspecto y de los informes, y rutinas de descarga más sofisticadas.

INTRODUCTION

The national PLANTS database contains taxonomic and distributional information for more than 36,000 taxa, all the terrestrial plants and lichens known to occur within the U.S. and its territories. PLANTS is a WWW-based system that can be reached at *http://plants.usda.gov.* NRCS biologists, planners, conservationists, and land stewards, and others in comparable positions in diverse state and federal agencies, are those traditionally served by the PLANTS system. However, the information provided is essential to monographers and those engaged in floristic studies, and PLANTS is widely used by them and others. The web site is consulted 60,000 times per month, and interest continues to grow.

NRCS's National Plant Data Center maintains the PLANTS database as part of its overall responsibility to provide plant information to the agency, its clients, cooperators, and the general public. The NPDC establishes cooperative agreements with universities, plant scientists, and others to help gather and improve plant information.

THE PLANTS DATABASE AND ITS USES

History of the Checklist

PLANTS evolved from the *National List of Scientific Plant Names*, which was first published by the Soil Conservation Service (SCS) in 1971. In 1978 SCS—which became NRCS several years ago—began revising this list in cooperation with the Department of Botany at the Smithsonian Institution, an effort which culminated in a new NLSPN in 1982 (USDA, SCS 1982). This two-volume document was the source for the alphanumeric plant symbols used by SCS in preparation of technical guides, handbooks, and soil surveys. The standardized plant names and symbols became essential for collecting, processing, and storing and retrieving the natural resource information used in all SCS's automatic data processing systems, and later in its computerized resource management tools, some of which are available via the web address for PLANTS.

By the late 1980's the 1982 NLSPN was becoming outdated, and during the revision that began in 1989, it was decided to transform the NLSPN to a new system called PLANTS (Plant List of Accepted Nomenclature, Taxonomy, & Symbols). At this time the list was updated and expanded, and, except in cases of duplication and incorrect spelling of the scientific name, this first PLANTS checklist retained the traditional plant symbols from the NLSPN. In 1995 PLANTS moved to the World Wide Web.

PLANTS: Checklist coverage

The national PLANTS database is a synonymized checklist for all vascular and non-vascular plants occurring in North America or the United States and its territories. Coverage varies somewhat between groups, thus the checklist now covers America north of Mexico for all ancient terrestrial plant groups, and covers the United States and its Caribbean territories for the vascular plants (Table 1). The overwhelming majority (84%) of the plants is vascular, and lichens—admittedly only marginally considered to be plants—represent the next most diverse group. Perhaps surprisingly, at nearly 5:1 the ratio of synonyms to accepted names is highest in the hornworts. Fortunately this ratio does not obtain in the vascular plants and other larger groups!

Vascular plants

The vascular plant checklist and nomenclature is developed through a cooperative agreement with the Biota of North America Program (BONAP), under the direction of John Kartesz. BONAP's comprehensive nomenclatural files have been assembled and revised over the course of two decades, and are based on 1) compilation of botanical literature; 2) assessment of the vascular plant collection available at the University of North Carolina at Chapel Hill and other institutions; 3) specialists' review; and 4) field data collected over the past 25 years. PLANTS contains state distributions for all states and Caribbean territories, and will soon have county distributions for 21 states (Table 2); both data sets are derived from BONAP data. For each of these occurrences, BONAP has documented either a bibliographical reference or voucher specimen data or both.

Lichens

Robert Egan (U. Nebraska at Omaha) and Theodore Esslinger (North Dakota State University) assembled the lichen checklist in PLANTS, which is based upon the North American lichen checklist published periodically in the *Bryologist* (Esslinger and Egan, 1995). The synonymized lichen checklist has been revised as of early 1999, and in cooperation with George Staples and his staff at the Bishop Museum in Honolulu, will soon be expanded to include Hawai'i. Roger Rosentreter of the Bureau of Land Management provided some of our lichen common names from an unpublished manuscript.

Mosses

Marshall Crosby of the Missouri Botanical Garden is currently updating our moss checklist, in association with botanists working on bryophytes for the *Flora of North America*. Lewis Anderson and Molly McMullen of Duke University developed the original list. The mosses in PLANTS are those occurring in continental North America, north of the Mexico border, though we soon expect to incorporate Hawai'ian mosses.

Liverworts and hornworts

Ray Stotler and Barbara Crandall-Stotler of Southern Illinois University developed these checklists. The hornwort nomenclature is current and complete, the liverwort data less so. We expect a comprehensive and fully synonymized liverwort checklist to be available through our cooperative agreement with Drs. Stotler and Crandall-Stotler sometime in the year 2000. The current list of liverworts and hornworts in PLANTS covers continental North America, north of the Mexico border. Hawaii hepatics will be integrated soon.

TABLE 1. Major data elements in the national PLANTS system. North America refers to the area north of Mexico.

Life form	Number of taxa below genus Accepted Names	/ %	Synonyms	Contributor	Scope and completness of checklist	Type of information	Data were updated in:	Notes
Vascular Plants	30,670	47	34,034	John Kartesz	USA, PR, VI; inclusive	Synonymized checklist with distribution	1999	Pacific Basin territories to be added in 2000
Mosses	1,511	44	1,918	Marshall Crosby	North America; inclusive	Synonymized checklist	1993	Hawai'ian taxa to be added in 2000
Liverworts	171	45	208	Ray Stotler and Barbara Crandall-Stotler	North America; incomplete	Synonymized checklist	1997	Hawai'ian taxa to be added in 2000
Hornworts	23	18	108	Ray Stotler and Barbara Crandall-Stotler	North America; inclusive	Synonymized checklist	1997	Hawai'ian taxa to be added in 2000
Lichens	4090	63	2388	Robert Egan	North America; inclusive	Synonymized checklist	1999 1999	Hawai'ian taxa to be added in 2000

The data in PLANTS and its uses

The core of PLANTS is its synonymized checklist of U.S. and North American plants. Attached to this checklist backbone are various "attributes" of each taxon (Table 2), especially vernacular name, distribution (Fig. 1), and concise information about growth habit. Other data fields of special interest to applied biologists include wetland indicator status as provided by the National Wetland Inventory in its *National List of Vascular Plant Species that Occur in Wetlands* (US Fish and Wildlife Service, 1988), and federal noxious weed and endangered species status. Information can be downloaded for the whole nation or individual states in a limited variety of formats. Photographs are provided for more than 2,000 species, and wetland plants are especially well represented. Some images in PLANTS are copyrighted but most are in the public domain; they are free for scientific and educational uses. Commercial use of copyrighted slides is prohibited unless permission is obtained from the copyright holder.

Please cite the information in PLANTS as: USDA, NRCS 2000. The PLANTS database. Http://plants.usda.gov. National Plant Data Center, Baton Rouge, LA, USA.

Utility of PLANTS

PLANTS serves as a national taxonomic standard for plant names, and is widely used by other federal agencies to standardize plant information and ease data sharing. In addition to providing a common language for floristic endeavors and resource management efforts, PLANTS continues to underpin NRCS's many automated resource conservation tools and Farm Bill programs. Some of these tools are reachable via the PLANTS web page, and include the Ecological Site Inventory System (ESIS), VegSpec (a digital mechanism for selecting plants for soil conservation and natural community restoration), and the NRCS Plant Materials program, which develops and distributes plants for conservation uses.

Partnerships and integration with other systems

PLANTS is a partnership and has been mutually supported by dozens of organizations and governmental agencies through the years (Table 3). Some of these collaborations have centered on data processing and gathering, others on data sharing and integration into other biodiversity

TABLE 2. Information pertinent to floristic and monographic studies from the national PLANTS database.

Type of Information	Description or data field values	Notes and future plans
Scientific name	Complete hierarchy for each terminal taxon	
Vernacular name	One per taxon	Multiple common names will be permitted in the future
Symbol	Unique alphanumeric identifying symbol for each taxon	Symbol=first two letters of genus+first two letters of species+first letter of terminal infraspecific name+tiebreaking number (if needed)
Growth Habit	Grass-like, forb (herb), subshrub, shrub, stem succulent, vine, liana, tree	Maximum of three permitted, though this will be expanded in the future
Duration	Annual, perennial	
Origin	Native to North America, introduced	Native plants with introduced populations are treated as native
Distribution	State; county for CA and MI.	Distribution data for vascular plants only. County data for AR, CT, GA, IL, IA, KS, KY, ME, MA, MO, NY, NC, ND, RI, SC, SD, VT, VA, WV to be added in 1999.
National Wetland Inventory Status	NWI categories are used: obligate, wetland, and all permutations	Current taxonomy is linked to 1988 *National List of Vascular Plants that* Occur in Wetlands
Noxious Weed Status	Federal Noxious Weed List is available	Will add individual state noxious weed lists with all categories and definitions in 1999.
Threatened and Endangered Species Status	US Fish and Wildlife Service federal status is available	Will add individual state lists with all categories and definitions in 1999.
Photographs	For over 2,000 taxa	Wetland plants are especially well represented.

information systems. A key partner is NRCS's Information Technology Center in Ft. Collins, Colorado, which develops and maintains the PLANTS database software and communications with assistance from the USDA National Information Technology Center.

PLANTS data are upwardly integrated into a growing number of biodiversity indices, including the Integrated Taxonomic Information System (ITIS). ITIS is a collaboration among seven federal agencies that will eventually provide a taxonomic structure of all biota of North America. PLANTS contributes the plant and lichen portions of ITIS. In similar fashion, PLANTS is included in the International Organization of Plant Information's Global Plant Checklist, the first iteration of which focuses on the Northern Hemisphere. The PLANTS database and Flora Europaea are the foundation data sets for a new software tool called MABFlora that was developed at the University of California at Davis for inventorying Man and the Biosphere sites. Lastly, Species 2000, a global effort to index the world's known species, is integrating ITIS data (and, in turn, PLANTS) as a regional data set.

Future improvements in the PLANTS database

By the time this article is published, PLANTS' users will be aware of a new appearance and new features. Lichen and vascular plant nomenclature will be completely updated in 1999 (Table 2), and county-level data will be added for about 20 states. All reports will be redesigned. Database queries and download capabilities will be improved by providing two options, a simple name search based on scientific name, vernacular name, or genus, and a more advanced search which will allow customized queries based on multiple geographic and attribute data fields. Using the advanced search one will be able, for example, to select and download all the threatened and

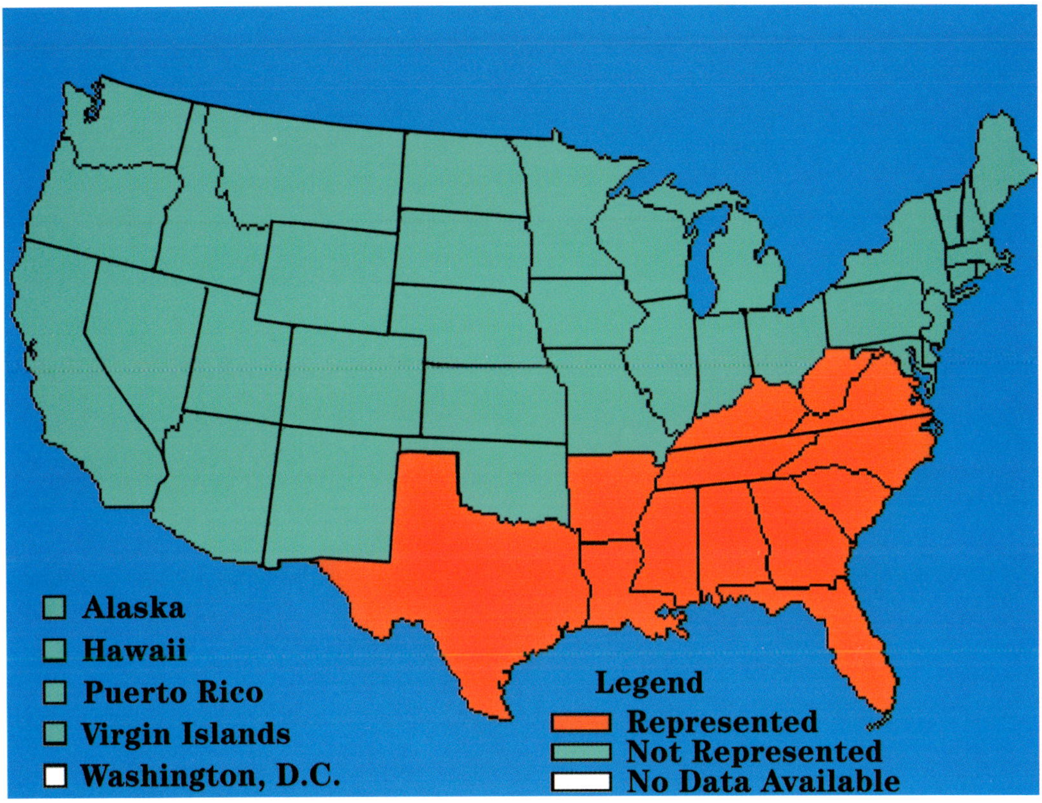

FIG. 1. National distribution of Carolina lily (*Lilium michauxii* Poir., LIMI) by state.

endangered herbs that are obligate wetland growers in the southeastern United States (i.e., US Fish and Wildlife wetland region 2). Various kind of information will be added or enhanced (Table 2), in particular state-level noxious weed status and plant endangerment status. The photo gallery will continue to grow, and will begin to incorporate links to other WWW sites that also display photographs of US plants. New links will be provided from species reports to plant fact sheets that describe preeminent conservation plants, and how to establish and grow them for restoration and habitat improvement programs. Other links will support new information about plants widely used by Native Americans and small-scale farmers.

Other NPDC activities especially relevant to research on the southeastern flora

Digital plant identification tool.—The NPDC is collaborating with Charles Allen at Northeast Louisiana University (NLU) to produce a digital key to the grasses and grass-like plants of Louisiana, a total of about 500 taxa. To some degree this will be applicable to a larger area of the Southeast.

Louisiana flora.—At the NLU herbarium, NPDC provided funding to Charles Allen to capture specimen label data to speed preparation of the final volume of the *Atlas of the Vascular Flora of Louisiana* (Thomas and Allen, 1998). This three-volume atlas is based on 350,000 specimens housed at all the major herbaria in Louisiana. NPDC also funded an effort to digitize collection label information at the Louisiana State Herbarium, which houses about 60,000 Louisiana collections. To date the Monocotyledons are digitized and Mark Mayfield has them available via the WWW at *http://LSB19.plbio.lsu.edu/herbdata.html.* Phil Hyatt at the Kisatchie National Forest has recently completed a database of that forest's herbarium, about 6,000 sheets. NPDC plans to combine these three data sets into a composite digital Atlas for Louisiana, and make this available in late 2000 via the national PLANTS database.

TABLE 3. Major partners with NRCS in developing the national PLANTS database.

Data Development Partners	Federal
Non-governmental	Agricultural Research Service
Alcorn State University	Animal and Plant Health Inspection Service
Bishop Museum of Honolulu	Bureau of Land Management
Flora of North America	Integrated Taxonomic Information System
Louisiana State University	National Oceanic and Atmospheric Administration
Missouri Botanical Garden	Smithsonian Institution
North Carolina A&T University	US Department of Defense
Northeast Louisiana University	US Environmental Protection Agency
Santa Barbara Botanic Garden	US Fish and Wildlife Service
Southern Illinois University-Carbondale	US Forest Service
University of Alabama	US Geological Survey
University of Colorado	**International Partners**
University of Guam	International Organization for Plant Information
University of Nebraska-Omaha & Lincoln	Botanisches Garten und Botanischer Museum
University of North Carolina, BONAP	Royal Botanic Garden, Edinburgh
Utah State University	Royal Botanic Garden, Kew
University of Wyoming	Royal Botanic Garden, Sydney
	International Legume Data Information System
	Species 2000

TABLE 4. Selected web-based botanic resources that are germane to the flora of the southeastern United States.

Title	World Wide Web Address	Notes
Atlases for SE floristic studies:		
Arkansas Biodiversity: The Vascular Flora	*http://www.csdl.tamu.edu/FLORA/arkansas/arkfamyy.htm*	Distribution maps at the county level for all Arkansas vascular plants.
Florida Vascular Plant Atlas	*http://www.usf.edu/~isb/projects/atlas/atlas.html*	Distribution maps at the county level for all 4,000+ Florida vascular plants.
Louisiana Grasses, Distributions and Diversity Maps	*http://www.csdl.tamu.edu/FLORA/lagrasses/lahome1.htm*	Distribution maps at the parish level, and state and parish checklists for all Louisiana grasses.
South Carolina Plant Atlas	*http://cricket.biol.sc.edu/herb/*	Distribution of all native and naturalized vascular plants of South Carolina.
Tennessee Vascular Plants Atlas	*http://www.bio.utk.edu/botany/herbarium/vascular/vascular.html*	Distribution maps of Tennessee vascular plants. Checklist also available.
Federal Government and international web sites with nomenclatural or floristic information. Only those referenced in the text are included:		
Germplasm Resources Information Network (GRIN) Taxonomy	*http://www.ars-grin.gov/npgs/tax/index.html*	Nomenclature for the accessions of the National Plant Germplasm System. About 35,000 taxa in 13,000 genera are included in GRIN taxonomy, including thousands that grow spontaneously in the United States.
Integrated Taxonomic Information System (ITIS)	*http://www.itis.usda.gov/itis*	Will eventually provide a complete taxonomic structure of all biota of North America.
International Organization of Plant Information Global Plant Checklist	*http://iopi.csu.edu.au/iopi*	PLANTS contributes the plants and lichens for the US part of IOPI.
Species 2000	*http://www.sp2000.org*	Species 2000 is a global effort to index the world's known species, and via ITIS, integrates PLANTS as a regional data set.

Web-based resources for southeastern US floristics that complement the PLANTS system

There are currently five digital atlases available via the WWW that offer range maps and checklists for one of the nine southeastern states (Table 4). These can be used in floristic and monographic studies to complement the national information provided by the PLANTS database, since all offer county-level distributions and some of the five cover plant habitats and their geomorphology.

REFERENCES

ESSLINGER, T.L. and R.S. EGAN. 1995. A sixth checklist of the lichen-forming, lichenicolous, and allied fungi of the continental United States and Canada. Bryologist 98:467–549.

THOMAS, R.D. and C.M. ALLEN. 1998. Atlas of the vascular flora of Louisiana. Volume III: Dicotyledons. Fabaceae–Zygophyllaceae. Louisiana Department of Wildlife and Fisheries Natural Heritage Program. Baton Rouge, LA.

USDA, NRCS. 1999. The PLANTS database. *http://plants.usda.gov.* National Plant Data Center, Baton Rouge, LA, USA.

USDA, SCS. 1982. National list of scientific plant names. Volumes 1 & 2.

US FISH AND WILDLIFE SERVICE. 1988. National list of vascular plant species that occur in wetlands. US Fish & Wildlife Service Biological Report 88(24).

A REVOLUTION IN COLLABORATIVE WORK PROCESSES I: THE FLORA OF NORTH AMERICA INTRANET

NANCY R. MORIN

The Arboretum at Flagstaff
4001 S. Woody Mountain Road
Flagstaff, AZ 86001, U.S.A.

Flora of North America is building on a long history of interest in the flora. It began in 1982 as a result of a meeting of plant taxonomists who were called together by Bob Kiger of the Hunt Institute for Botanical Documentation and Duncan Porter from Virginia Polytechnic Institute, who had both been involved in Flora North America in the 1960s, to discuss strategies for reviving the flora effort. In 1984 an editorial committee was assembled, and they developed the plan for the project, which was to have an organizational center at Missouri Botanical Garden and a Bibliographic center at the Hunt Institute. Funding was first received in 1987, and the project has been funded since then by grants from the National Science Foundation, many non-governmental foundations, private donors, and very significant in-kind contributions from participating institutions (Morin & Spellenberg 1993).

Flora of North America provides a unified framework for basic and applied research dealing with North American plants and plant products. Diverse kinds of information that have been carefully verified by project staff and reviewers are available from a single source in a uniform format, and differing taxonomic opinions from many widely used references are reconciled and related to an authoritative standard. Plant anatomists and morphologists, ecologists, phytochemists, evolutionary biologists, and molecular biologists in addition to taxonomists, all need accurate, complete information in order to circumscribe their studies appropriately, to select representative taxa for detailed investigations in large groups, and to interpret their results in relation to the massive body of data now available for many groups. Information in the database, particularly when linked with precise locality data found on herbarium specimens, can be analyzed to discover correlations among geographical distribution, co-occurrence with other organisms, genetic relationships, and morphological or chemical attributes. Flora of North America provides a gold mine of information for further research.

Flora of North America covers North America north of Mexico: the continental United States of America, including the Florida Keys and Aleutian Islands, Canada, Greenland, and St. Pierre and Miquelon islands, which together form a phytogeographically significant region (Thorne 1993) and occupy 21,461,119 square km. This area is tremendously diverse topographically, geologically, climatically, and vegetationally. About 20% of the flora of the area is endemic, and certain areas, such as California, Florida, and Texas, have many endemic taxa.

Flora of North America fosters scientific dialog by bringing together specialists on major groups of plants to exchange ideas, discuss taxonomic problems, come to agreement on taxonomic concepts, and decide on what and how attributes are to be handled. In many cases, the FNA treatment will synthesize the concepts developed by an entire generation of plant taxonomists. For example, many of the pteridophyte treatments in volume 2 were prepared by Warren H. Wagner and his students and associates. Their work has radically altered our view of North American pteridophytes, much of it previously available only in scattered research reports, often dealing with single species. The FNA treatment provides the only detailed synthesis of this large and important body of work, making it truly accessible to the general scientific community for the first time. FNA includes thorough documentation for the taxonomic treatments, including detailed, parallel descriptions and extensive literature citations, and thus provides the user with the means to investigate and evaluate the basis of the treatments.

The project began with some important philosophies. Its goal has been to gather the best

SIDA, Bot. Misc. 18: 61–65. 2000

possible information on the plants of North America and make that information available in useful ways. This is a simple goal that is enormously complex in its implementation.

Gathering the best possible information

One policy of FNA has been to invite **experts** to provide treatments. If there is an expert on a small complex of species, he or she will be asked to write the treatment for that group. As a result, FNA works with some 750 authors. It can be argued that it would be much more efficient to deal with fewer people, but, project participants believe that this approach gets the best information, and it seems likely that this is the only time that this approach will be possible because the expertise to do it will not be available in the future. The treatments are **reviewed** by more than 50 regional reviewers, and each treatment is reviewed taxonomically, in order to achieve another goal of the project, which is to present a comprehensive view of each taxon throughout its range. Since few authors have been able to do field work or examine specimens from throughout North America, expertise from reviewers has been tremendously important. The review process is not a matter of superficial checking; for FNA, reviewers are an integral part of the process of getting the best possible information. The FNA definition of the "best possible information" includes more than correct nomenclature, appropriate literature references, and accurate descriptions, distributions, illustrations, and discussion. It also includes the concepts of completeness, comparability, and internal consistency. In addition to taxonomic reviewers, FNA editors and staff play a very important role in assuring that these criteria for "best possible information" are met.

Authors, reviewers, editors, and staff all need easy access to information resources in order to prepare, review, and edit treatments. Few taxonomists have the good fortune to be located near a comprehensive botanical library or a major herbarium. Flora of North America attempts to alert participants to important resources. Few participants have the time or money to travel to these resources, however, and, consequently, the most useful and most practical way to provide access to information is via the Internet.

Making the information available in useful ways

Flora of North America, from its inception, has had as a goal both paper and electronic publication of its information, the electronic form being a fully relational database. The content, in order to be most useful, is to have a minimum of technical terminology, and to be consistent and comparable. Additional information, such as economic uses, weediness, toxicity, and conservation status are included, as well as maps and illustrations, in order to serve a broader audience. Perhaps one of the most important considerations under the "usefulness" criterion is that the information be provided in a timely manner. The editorial committee, when the project started in 1984, considered this to be "within our professional lifetimes." Our reasoning for this was not only personal interest in seeing the project to completion but also realization that good, comprehensive information on the plants of North America is urgently needed. Habitats are still being lost at an alarming rate; and one in three species of plants in the U.S. is under threat of extinction. Government and private organizations are attempting to manage developed and natural landscapes and to protect species as best they can, but they must have good data to do this successfully.

Print publication of a work this large, with an estimated 30 volumes, is in itself an enormous task The three volumes that have been published to date have each sold more than 5000 copies, confirming the public's need for the information in book form. But because printed information is static—that is, it cannot be changed or corrected except in subsequent printings—the editors and staff feel a particular obligation to assure that the published information is correct. As a consequence, preparation of sets of treatments for publication is an intense and completely engrossing activity for the editors and staff.

TENSIONS

These many aspirations of the project result in conflicting needs, skills, and priorities. An underlying fact of life that influences the project's progress is limitation of time. Authors and reviewers and editors all work as volunteers. Virtually all of them have full-time jobs, active research programs, families, and other things to do in addition to FNA. They have very little time, and it is incumbent on the FNA project to use that time effectively, and to cause a minimum of frustration.

If we are to complete FNA "in our professional lifetimes," work has to be done on most families simultaneously. This is possible at the authoring stage, since most specialists are writing treatments for only one or a few groups. Once a treatment has been submitted, however, the project structure becomes a limiting factor.

Even though FNA incorporated the use of computer technology from the beginning as a way of delivering final information products, until recently processing of manuscripts has followed the most traditional patterns of journal and book preparation. Hard-copy manuscripts are received (now virtually always with diskette, and if not then scanned to provide an electronic form). Paper copies of the manuscript are mailed to reviewers. A reviewer makes comments on a paper copy and mails it to either the editor (if a taxonomic review) or the regional coordinator for his or her area. Regional coordinators accumulate marked manuscripts, evaluate comments and then transcribe them on to a single copy, which is sent to the taxon editor. The taxon editor evaluates all comments, transcribes them on to a master copy, and mails that to the author with recommendations for revisions. For widespread taxa, more than 50 people will be involved in the review process; for genera with many species, an individual manuscript may be several hundred pages long. The time and expense required for the simple physical handling of all of this paper is enormous. Even if time and money were not limited, it would not be a productive use of them. But as it is, time and money *are* limited. Priorities for the FNA project generally relate to the next volume to be published, although special attention is given to treatments by specialists with their own rate-limiting issues, such as age. Treatments that will not be published in paper form in the next several years inevitably, and appropriately, receive lower priority for processing. Taxon editors are daunted by the piles of paper. Regional reviewers receive large numbers of treatments sporadically. The most distressing aspect of this problem is the loss of intellectual momentum. A regional reviewer may make an important suggestion that doesn't reach the author until many months later (and then, since it has been transcribed and combined by other people in the interim, it may not fully reflect the reviewer's intention). If the author wishes to question or explore that comment further, it may take another several months before his or her question reaches the reviewer, who by that time can probably barely remember what the idea was. One of the most remarkable features of FNA is that authors, reviewers, and editors, have managed to be so patient and good-spirited about producing a quality product despite these limitations.

MANAGING MANUSCRIPTS ON THE INTERNET

An almost magical solution to this problem seems to be offered by the Internet. If submission, review, and editing can all be done using the Internet as a virtual office, then editorial teams (authors, reviewers, and editors) focused on specific groups can work together directly, without the time-intensive logistical support of the office staff. Furthermore, theoretically it should be possible for authors, reviewers, and editors to have nearly real-time dialogs in which they all can view each other's comments and respond directly.

One of the areas in which we think this will make the most difference is in the preparation of distribution maps. Distributions have been the most challenging aspects of preparing and editing FNA treatments. As we see in this symposium, depth of coverage of local areas is best in quite small herbaria, from which authors may not have known to request specimens. In addition, many authors work at institutions for which borrowing and returning specimens can be a major drain of resources. Similarly, many authors do not have access to all the recent atlases. Regional reviewers are knowledgeable about the plants from their area represented in herbaria, about the literature reports for their flora, and about local field knowledge. The review process helps bring to light misidentifications, differences in understanding of circumscriptions, and places where the treatment is ambiguous. Direct dialog, ideally taking place as participants view each other's concepts, is surely the best way to achieve consensus and mutual understanding.

The greatest challenge to making the most of the opportunity presented by the Internet is managing the process. To throw the entire FNA process open to all participants would assure pandemonium and grid-lock. The traditional processes may be tedious and time-consuming but they are orderly, and they include archival records of what was said by whom, and when. FNA archives almost all marked manuscripts—a historian in the future will be able to piece together the thought processes and professional evolution of many of the attendants of this symposium.

For many years, FNA has had a database accessible to all editors that tracks progress of each treatment, but it doesn't give explicit guidance to editors as to what needs to be done. The Center for Botanical Informatics (CBI) at Missouri Botanical Garden worked with FNA staff, editors, and other participants to develop protocols and identify needs for using the Internet as a virtual office. The first phase of this development focused on making the published treatments and information resources available on the Internet. The next phase is to make manuscripts in review and revision available.

CBI viewed pieces of information as information entities that can be interpreted in several ways so that the needs of any particular participant are met. Their technical term for this is "boundary objects." The challenge they saw themselves as addressing is "Managing Cognitive Overload"—which is probably a very apt phrase.

For FNA, the system must accommodate the needs of a Taxon Editor, other editors, reviewers, and authors. The taxon editor needs reports to monitor overall programs and must be able to add or delete assignments. The author, in contrast, needs to see only the status of his or her own manuscript. Each reviewer may need to be alerted to availability of a number of treatments. Our strategy for dealing with this is to give each participant dynamically constructed, customized home pages that allow viewing information according to the role a person plays in the project.

In other words, one manuscript may be viewed by the author, who sees only this, by the taxon editor, who sees this and all other manuscripts for which he or she is responsible, and by the regional reviewer, who also sees a different suite of treatments. The system knows who you are and what role or roles you play based on your login information. One additional, very important benefit of using the Internet as a virtual office, is that ancillary information—such as lists of specimens seen, type information, phylogenetic analyses, etc.—are available to those who are working with the treatment itself. The system tells you what is available, what state it is in, what your own next action should be, and what your deadlines are. A simple approach to this is to make manuscripts available using File Transfer Protocol (FTP), with access to individual manuscripts restricted based on the role of the person accessing them.

A very important point is that everyone sees custom views of the *same* information. We are all singing from the same page of the choir book. And the information, which was presented in a general table before, is now put in the context of processes and editorial tasks.

The concepts used, while intuitively obvious, are computationally complex. Furthermore, their implementation is based on assumptions about what really happens in the project, how a particular individual thinks, learns, and works. CBI has conducted planning meetings and focus groups. Probably the first barrier that has been met is the problem of getting editors used to checking the Internet site regularly and to working in that environment.

PUTTING THE DATA IN A FORM THAT CAN BE ANALYZED

Another goal of the project has been to make data from treatments available in a relational database. The ideal solution seemed to be to use DELTA, and some authors are doing this. For most authors, however, preparation of a character and character state matrix sufficiently detailed and flexible for their group involves learning a new method and investing time disproportionate to the advantage gained. Consequently, FNA deals mostly with treatments prepared in a traditional text form. Editing such manuscripts for consistency and comparability is labor intensive, especially for large genera or families with many genera, and technical editing to move characters and descriptors into correct order is also very time-consuming. Our solution has been to parse the treatments using keys such as typeface (e.g., bolding, italics) and punctuation, and to port the information into a spreadsheet format. The resulting columns can be checked for presence or absence of information and for internal consistency in use of terminology. They can be moved and edited en mass, a great time saver. Segments of descriptions can be searched separately, although this still is only a coarse solution.

A criticism that is being leveled at taxonomists in general is that they capture the same information again and again. Treatments have been written for local floras or regional floras, or for the plant as it occurs in another part of the world, and yet descriptions are written new each time for each project.

Alfredo Sanchez, of the Universidad de las Americas-Puebla, working with CBI, has taken a novel approach to solving both of these problems—how to get the descriptions into a database format and how to take advantage of work that has already been done. He suggests using User Agents, intelligent programs that can discern a pattern in information as it is being developed. For instance, if you described leaf shape in the first three descriptions, when you got to the fourth description it would prompt you for leaf shape. And as the structure develops, the information is parsed into appropriate fields. Sanchez used techniques found in digital libraries to go one step further. When you begin to describe a taxon, the user agent checks all other files, tells you what it finds, and asks if you want to use that information again. If the system had a description for *Triodanis perfoliata* from Flora Mesoamericana and I began to describe it for Flora of North America, the system would bring up the existing description and allow me to alter it as appropriate. These programs are just in the development stage, but if he is successful, this would be a tremendous help.

CONCLUSION

We are excited about the progress that is being made for using new techniques to assist participants as they work on Flora of North America. Hundreds of botanists have already devoted a great deal of time and effort to the project, are working on it, or will work on it. Not only do we selfishly want to finish this project in our lifetimes but we have a deep obligation to society to make this important expert information available for conservation and resource management.

But we think another reason to be excited is that, if these systems can be successfully implemented, they can be useful to all of the large-scale collaborative projects in any field.

REFERENCES

MORIN, N.R. and R.W. SPELLENBERG. 1993. History of the Flora of North America Project. In: Flora of North America Editorial Committee, eds. Flora of North America Editorial Committee, eds. Flora of North America North of Mexico. Volume 1: Introduction. New York. Pp. 3–12.

THORNE, R. 1993. Phytogeography of North America north of Mexico. In: Flora of North America Editorial Committee, eds. Flora of North America North of Mexico. New York. Volume 1, pp. 132–153.

A REVOLUTION IN COLLABORATIVE WORK PROCESSES II: STRATEGIES FOR TREATING THE BIG FAMILIES[1]

THEODORE M. BARKLEY[2]

Division of Biology
Kansas State University
Manhattan, KS 66506, U.S.A.

Floristic studies have changed notably since the present generation of botanists came of age. In the not-so-distant past, a floristic study was expected to be a competent reflection of what plants occur in a region, and it included keys and descriptions to aid in their identification. A floristic study still includes those items, but botanists and the consumers of botanical information now expect much more. A flora for the 21st century, i.e., a flora by our generation of botanists, also must include a high degree of accuracy for distributional data, phenology, and biological information and must summarize the current taxonomic understandings for the various groups.

The amount of current information relevant to floristics is utterly terrifying, as even the most casual peruser of the literature knows. It is beyond the ken of any one person, no matter how dedicated, to have a full grasp of current thought in systematics, and so a modern floristic venture includes a component of computer-driven information management, or "informatics." Several years ago, the annual fall symposium at the Missouri Botanical Garden was titled "Knowledge Brokering... the Techniques of Synthesis," the gist of which was that electronic information management was on the horizon. Now it is here, and we simply cannot address all of the expectations of modern floristics without "informatics."

These notions were important in the conception of the current "Flora of North America" project (FNA), which seeks to be a summary compendium of the continental flora north of Mexico. The administration of the FNA originally thought of the product as a dozen bound volumes, but as the first two volumes were readied for print, the project was expanded to include 26 volumes for vascular plants and another three for bryophytes. Also, a clearer idea now exists of the FNA on-line component, which is an ever-current version of the FNA that is accessible via the FNA website.

The FNA Organization was soon aware that the project would need the collaboration from professionals in the field of information management and would require a spirit of cooperation among all involved. Systematic botanists traditionally have worked more or less alone, so a challenge for the FNA has been to build a program of comfortable cooperation among the botanists, informatics specialists, and the intended users of the product.

Background.—The contributing authors are clearly the most important parts of the FNA project, for their expertise forms the viscera of what the project intends to communicate. To facilitate the project, the FNA has an administrative group, the FNA Central office, or FNAC, located at the Missouri Botanical Garden in St. Louis. The FNAC's responsibilities include managing and editing manuscripts, arranging for preparation of maps and illustrations, and achieving editorial conformity among the products. Each natural group of plants (family or portion of a family) is overseen by a Taxon Editor, who works closely with the authors. Manuscripts are reviewed for regional accuracy by several Regional Coordinators, working with regional reviewers who test the manuscripts for accuracy and verify distributional and ecological information. Subsets of the administrative group work on maps, illustrations, nomenclature, and bibliographic aspects. A Contributors Guide (Flora of North America Organization 1996) presents the principles and

[1]This is contribution no. 98-453-J from the Kansas Agricultural Experiment Station, Manhattan.

[2]Present Address: Botanical Research Institute of Texas, 509 Pecan Street, Fort Worth, TX 76102-4060, U.S.A. email barkley@brit.org

SIDA, Bot. Misc. 18: 67–71. 2000

matters of style for the manuscripts and the products. This guide makes several assumptions, including that authors will account for all of the relevant names that have been used in their group in the FNA region. All descriptions will be strictly parallel, in accord with certain conventions, and all keys will be dichotomous. Moreover, the guide assumes that each author will essentially exhaust the relevant library and herbarium resources in preparing his or her contributions. A noteworthy point is that except for the small FNAC staff and some informatics specialists, the FNA rests on volunteer participants. The authors, taxon editors, regional coordinators, and reviewers are mostly academics or institutional researchers whose home institutions either generously or grudgingly permit some staff time and facilities to be committed to the FNA. Regardless of the good intentions by authors and others, few of the participants in the FNA have the luxury of complete dedication to the project, and this has implications in both the planning and the execution of the venture.

The Large Families.—It is one thing for a group of authors and a taxon editor to prepare the treatment for a family of a few genera, but quite another matter to prepare a treatment for the larger families with dozens of genera and complex problems in taxonomic concepts and nomenclature. Current estimates are that the treatments for the grasses and the legumes will each occupy two volumes and the treatment of the composites will fill three volumes. What follows is a discussion of the FNA program for the Compositae, but it is equally relevant to the programs for the grasses or the legumes and, therefore, any large family.

By a conservative count, the Asteraceae in the flora of North America north of Mexico encompasses some 2500 species scattered among nearly 400 genera. The past quarter century has seen great revisions in tribal and generic concepts, and most students of the family agree that the newer and often narrower concepts are here to stay (Bremer 1994; Barkley et al. 1996; Hind & Beentje 1996). The FNA must assess these studies and assimilate the new concepts into a modern, useable flora. In order to do this, the Editorial Committee of the FNA (the administrative board for the project) created a Compositae Editorial Committee, or "CompEd," consisting of John Strother of the University of California-Berkeley, Luc Brouillet of the Université de Montreal, and Theodore Barkley of Kansas State University. The three CompEd members prepared tentative lists of the tribes and the genera to be included in the project. A great aid in this has been the "List of Names to be Accounted For" that is maintained in a state of constant editorial development at the FNAC, and is available through the FNA website. This list is simply a compilation of all referable names from some 25 regional North American floras, and the notion is that if a name appears on that list, it must be accounted for somewhere in the FNA. Then the CompEd sets about soliciting authors to contribute treatments to the project. At about the same time, volume 2 of the FNA, which included the ferns and fern allies, was approaching publication. Fern systematics have undergone considerable revision in the recent past, and advice from the taxon editors of the ferns was sought in helping the CompEd to integrate the many new notions in the taxonomy of the composites.

As a result of deliberations and advice, the CompEd generated three policy matters that have been of great utility:

1.—Tribal circumscriptions were recognized in accord with Kare Bremer (1994), whose summary compendium of the family is both contemporary and accessible to all. Editorial responsibility for the tribes was divided as follows: L. Brouillet: Lactuceae, with 44 genera and 235 species, Mutisieae 7/15, Arctotidae 3/3, Calenduleae 3/4, Cardueae 14/120, Gnaphalieae 14/199, Inuleae, s.str., 5/7, Plucheeae 4/15, Anthemideae 20/100, and Astereae 64/891; J. Strother: Heliantheae, s.l., 139/740; T. Barkley: Senecioneae 31/156, Vernonieae 9/30, and Eupatorieae 30/150. The totals recognized at the start of the project are 386 genera and 2564 species, but naturally, all of these figures are subject to revision as the project progresses. In general, the three CompEd members serve as taxon editors for each of their groups, but with some modification. Strother is an experienced botanical editor who prefers editing to the negotiations with authors, so he will do a larger share of editing than just the Heliantheae. Barkley, rather than Strother, works directly with the authors in the Heliantheae.

2.—The CompEd members prepared "working descriptions" for each of the genera of composites to be accounted for in the FNA and appended a list of names relevant to each of these genera. The CompEd thus has a means to account for all taxa, which will help ensure that no taxon is omitted accidentally or that no entity is treated under two names. These working descriptions then are supplied to the authors as starting points for their work. An author, of course, may prefer different generic concepts and is free to adopt them, in consultation with the CompEd.

3.—A special addendum to the FNA contributor's guide was prepared largely by J. Strother to account for matters of particular relevance to the Compositae. Included are precise definitions of terms for describing the composite inflorescence and capitulescence, a family description with terms used in their proper contexts, and model treatments of several genera in the Heliantheae, s.l. The CompEd addendum is bound with the version of FNA Guide for Contributors that is given to authors of composite groups.

Authors to contribute treatments of Compositae were sought widely, and at present some 90 authors have agreed formally or have given verbal assent to contribute. Most genera have been assigned to authors and, as this is written, only some 35 genera are still "orphans." The CompEd retains the right to negotiate with authors, i.e., to assign treatments to authors and to withdraw assignments as dictated by circumstance, so if an author cannot deliver a treatment, the CompEd may pass the assignment to another author. Furthermore, CompEd members are prepared to write some treatments of orphan groups, if need be. It is assumed by all that the CompEd oversees the preparation of the Compositae treatments and is responsible for delivery of the product to the FNAC.

SYNTHESIS OF THE TREATMENTS

Assimilating all of the information and producing a useful product "by hand" are theoretically possible, however, the FNA in general and the work of the CompEd in particular are tailor-made test-beds for the application of informatics procedures. The encompassing nature of the project and the depth expected by the consumers of botanical information rather preclude a simple, superficial compilation. Computer-driven schemes for information management provide the prospect of the expected depth for the FNA, and all who are associated with the project have understood this, at least in some way. It is fair to note that the complexities of applying informatics techniques were but dimly grasped as the FNA began its work. The participating botanists were not completely articulate in communicating their exact needs to the computer scientists, and vice versa, and so there has been a mild culture-clash. Regardless, four matters have come from this:

1.—The FNA will be published in a hard-copy format.

2.—An accessible, on-line version of the FNA will be maintained in a state of constant revision. A CD-ROM version has been discussed, but at present that aspect is not active.

3.—The FNA, in collaboration with the Center for Botanical Informatics (CBI) of the Missouri Botanical Garden, will have a scheme for the easy transfer of manuscripts among the various players, e.g., authors, taxon editors, regional reviewers, and manuscripts in-progress will be accessible to FNA operatives all along the way. (This is the Collaborative Publishing Services, or CPS, that was spoken of by our colleague, Dr. Nancy Morin, Convening Editor of the FNA). The CPS is just coming into operation as this is written, but clearly it is a powerful editorial tool.

4.—A scheme for authors to compose keys and descriptions via computer-assisted methods is all but promised by the developments in computer science. DELTA is a well-known system for doing so, and it offers much, but some thoughtful planning will be needed to bring a DELTA format into operational conformity for the FNA-CompEd applications.

At present, a taxon-driven "backbone" is being compiled for the composites. This is a list of all tribes, subtribes, and genera recognized at the start by the CompEd, along with their places of publication. Epithets for recognized species are listed alphabetically, and then they are re-aligned in the taxonomic sequences of the contributing authors, as the authors' manuscripts become available. This "backbone" is of inestimable value for the planning of illustrations and placement of maps and for giving any author or editor an overview of the status of any treatment. Computer technology assures the ease of revising the backbone as new manuscripts arrive or as revisions are made. The Compositae backbone is accessible on the FNA website.

MATTERS FOR CONSIDERATION

The CompEd and all of the FNA organization must be ever mindful that the botanical expertise that forms this project comes from specialists who are mostly volunteers, and that an important purpose of the project is to incorporate what they know about the plants. The CompEd has experienced some reticence by a few authors to buy into the electronic information age. As a consequence, the CompEd and the FNA will need to make special efforts to assure authors that their contributions are greatly valued, even if the author is uncomfortable with computer applications; it is the botany, and not the conformity to an information management scheme, that is of ultimate interest. By extension, the CompEd must be prepared to supply gentle and respectful instruction in the vagaries and byways of the FNA's electronic information management schemes to any of the authors who request it.

Little has been said here about the distribution maps prepared for the FNA. Nearly every participant in the project, from author to editor to reviewer, has an opinion about the quality of the maps, but most agree that the small maps that are currently a part of the FNA style are much better than no maps at all. The message to all authors, however, is to make the maps as accurate as possible, and that means consulting all of the published maps, atlases, and herbaria as is reasonably convenient. We see in the future the development of very accurate and information-rich maps that are available on-line and that can be updated as new information becomes valuable. The CompEd is acutely aware of the importance of good distributional information, and authors are urged to generate as much hard data as comfortably possible.

SUMMARY

Preparation of the FNA treatments for the legumes, grasses, and composites is off-and-running. From the perspective of the Compositae Editorial Committee, the most important single item that has been done for the authors is supplying each of them with a tentative working-description of the genus (or genera) that he/she has agreed to prepare for the FNA. This tentative description, plus the list of names to be accounted for, circumscribes each author's responsibility. Two procedural matters are of importance: (a) authors must be contacted regularly so that each author retains a sense of participation in a successful, on-going project, and (b) each author must be comfortable in the FNA even if he or she is not fully operational in the computer-driven schemes for information management.

ADDENDUM

Since this paper was presented in 1998, some notable changes have occurred in the FNA organization. The FNA Central Office at the Missouri Botanical Garden has been reduced in size, and much of the work of preparing the treatments has been dispersed among several Editorial Centers across the FNA region. The functions of the CompEd are now located at the Editorial Center at BRIT. Barkley has retired from Kansas State University and relocated to BRIT to assume a leadership role in the BRIT Editorial Center. It is expected that the diffusion of administrative and editorial efforts among several Editorial Centers will increase the efficiency of moving the FNA forward, toward completion. There has been a reconsideration of the ways that computer-based technologies are to be used in the project, and the work formerly done by the CBI is now in the hands of the FNA organization itself. Potential contributors and participants in the FNA are assured that the project is alive and well, and that the mood is upbeat.

REFERENCES

BREMER, K. 1994. Asteraceae: Cladistics & classification. Timber Press, Portland, OR.

BARKLEY, T.M., B.L. CLARK, and A.M. FUNSTON. 1996. The segregate genera of *Senecio*, s.l., and *Cacalia*, s.l., (Asteraceae: Senecioneae) in Mexico and Central America. In: D.J.N. Hind and H.J. Beentje, eds. Proceedings of the International Compositae Conference, Kew, 1994. Royal Botanic Gardens, Kew.

HIND, D.J.N. and H.J. BEENTJE (eds.). 1996. Compositae: Systematics. Vol 1, Proceedings of the International Compositae Conference, Kew, 1994. Royal Botanic Gardens, Kew.

FLORA OF NORTH AMERICA ORGANIZATION. 1996. Guide for contributors (Compositae version). Flora of North America, Missouri Botanical Garden, Saint Louis. Flora of North America website: http://fna.org.

USING GEOGRAPHIC INFORMATION SYSTEMS IN FLORISTIC STUDIES

LEILA M. SHULTZ

College of Natural Resources
Utah State University
Logan, UT 84322-5215, U.S.A.
and

Harvard University Herbaria
22 Divinity Avenue
Cambridge, MA 02128, U.S.A.
email: <shultz@oeb.harvard.edu>

ABSTRACT

Geographic information systems (GIS) provide a visual, intuitive interface for the integration of complex data sets and are ideally suited for relaying information about spatially distributed objects such as plants. Because systems available before the mid-1990's were prohibitively expensive for most systematists, early developers of herbarium database systems did not consider the possibility of capturing locality data from specimens without creating a comprehensive database. This paper discusses a system that takes advantage of simple technology to integrate data from published records, distribution maps, and specimen-based data systems. The working floristic geographic information system for Utah integrates contemporary voucher records with geo-spatial data from several sources, including specimen-based distribution maps from a published atlas of plants. Elements of the system were developed in response to growing environmental and conservation needs in the state of Utah and provide a model for rapid assessments of biodiversity. Our floristic geographic information system is being used to identify areas of conservation concern, develop a model to predict the distribution of rare plants, and analyze patterns of biodiversity. Inexpensive GIS technology allows floristic botanists to use analytical systems available for personal computers without becoming enslaved in the development of massive databases. A simplified means for capturing geo-spatial data from herbarium specimens without creating a detailed database of label information is discussed.

KEY WORDS: geographic information system (GIS), floristics, Utah

RESUMEN

Los sistemas de información geográfica (GIS) ofrecen una interfaz visual e intuitiva para la integración de conjuntos de datos complejos, y están adaptados para transmitir información de objetos distribuidos espacialmente como las plantas. Debido a que los sistemas disponibles antes de la mitad de los 90 eran demasiado caros para la mayoría de los botánicos sistemáticos, los primeros promotores de sistemas de bases de datos de herbarios no consideraron la posibilidad de capturar datos de las localidades sin crear una base de datos comprensible. Este trabajo discute un sistema que aprovecha la tecnología para integrar datos de registros publicados, mapas de distribución, y sistemas de datos basados en especímenes. El sistema de información geográfica para trabajar en florística en Utah integra registros de testigos contemporáneos con datos geo-espaciales de diversa procedencia, incluyendo mapas de distribución basados en especímenes de un atlas de plantas publicado. Los elementos del sistema fueron desarrollados en respuesta a las necesidades crecientes en medio ambiente y conservación en el estado de Utah y ofrece un modelo para cálculos rápidos de biodiversidad. Nuestro sistema de información geográfico florístico se usa para identificar áreas para conservación, desarrollar un modelo para predecir la distribución de plantas raras, y analizar patrones de biodiversidad. La tecnología barata GIS permite a los botánicos floristas usar sistemas analíticos disponibles para ordenadores personales sin esclavizarse con el desarrollo de bases de datos masivas. Se discute un método simplificado para capturar datos geo-espaciales de especímenes de herbario sin tener que crear una base de datos detallada con la información de las etiquetas.

INTRODUCTION

Systematists work to discover, organize, and interpret biological diversity. As the discipline that endeavors to describe organisms, document their patterns of distribution, and unravel their genealogies, systematics is in obvious need efficient, easily shared systems of information. Finding common ground for communication is not easy. In the past two decades, different systems for organizing biological data have proliferated and most institutions are locked into using one format or another. Even the simplest of data systems, such as a catalogue of specimen localities, can be so complex and take so many different forms that interpreters of patterns of diversity become slaves to the task of organizing and editing data files.

Floristic studies are an integral part of plant systematics, providing specific information about organisms and their habitats. The floristic botanist documents patterns of variation in the field. Ideally, this botanist also works to describe and analyze patterns of biological diversity in the context of geographic and phylogenetic relationships. Working at the landscape level in a discipline founded on attention to detail often leaves little time for analysis. It is no wonder that few people are willing or able to tackle the job of the floristic botanist. Yet for the first time, tools of analysis that facilitate the analysis of massive amounts of complex data are commercially available. How, then, do we approach the problem of capturing the massive amounts of data without becoming slaves to data management. How, for instance, do we assimilate existing information for a specific region such as the southeastern United States?

The problem botanists face in North America is not of too little data, but data of varying quality and quantity. We have massive amounts of historical data (more than we have been able to assimilate and manage) and very little contemporary information about plant distributions. What we already know (i.e. what has already been documented by an estimated 60 million voucher specimens in North America) about the distribution of a species usually lies dormant in hundreds of different herbarium collections.

The case for efficient modes of information capture and exchange is indisputable, yet in North America there is no geographic standard for reporting the distribution of flora and fauna. Given the number of different management programs used in the United States, we may never have a uniform system. The absence of a unified national information system for biodiversity continues to hamper research efforts and national assessments of conservation needs. It is my position in this paper that we must identify the *minimum* amount of information needed for an analysis of distribution patterns as opposed to the maximum. Given the massive amounts of information housed in different institutions, we should be asking what data is essential. For purposes of identifying and examining plant diversity, the essential data is simply *what grows where*–the name of the organism and where it is found. My premise in this paper is that there are direct and efficient ways to incorporate the information provided by specialists and housed in herbarium collections without embarking on the creation of large and complex data sets.

Commercially available geographic information systems (GIS) are immensely useful in the task of organizing complex data and analyzing patterns of biodiversity. Stimulated by some of the pioneering work by Scott et al. (1987) in developing geographic systems to analyze gaps in areas where biodiversity is protected, a growing body of work demonstrates the utility of using these systems to identify patterns of diversity and prioritize areas for conservation (Beaman et al. 1998; Kareiva 1993; Rebelo & Tansley 1992; Rebelo 1994; Scott et al. 1993; Shultz et al. 1998a; Shultz 1998b). But developing a simple strategy for capturing biologically relevant data remains a daunting task. While the importance of a geographic interface with herbarium management systems was recognized early in the development of herbarium databases, most herbarium databases in the U.S. can provide little more than coarse resolution data such as state or county distributions (Rhoads & Thompson 1992).

As the cost of commercially available GIS programs continues to decline, more and more researchers will find the programs attractive. The systems I discuss show how historical, coarse-resolution geographic locations from herbarium specimens were joined with site-specific studies in Utah to further conservation efforts and stimulate new directions in research. By mapping species distributions without creating a specimen database, we were able to develop a comprehensive and useful system in a short period of time.

DEVELOPING AND USING A BOTANICAL GEOGRAPHIC INFORMATION SYSTEM

One of the premises of developing a meaningful geographic information system is that the data must be stated in a format that can be spatially referenced. Unfortunately, we have never used a uniform grid system to report the distribution of species in North America. Most labels for herbarium specimens provide state or province and county distribution statements, accompanied by general locality. In the eastern U.S., where counties are small and collectors have been active, county distributions provide a reasonable portrayal of where species occur. Contrast this with the sparsely-collected western U.S. where counties are larger than many eastern states, and one sees the limited utility of county-based distribution statements. The project to map locations of herbarium collection of Utah plants was meant to rectify this limitation. Knowing that most pre-1980 herbarium collections provide vague locality information, we looked for a practical method. Following the grid-based example of successful mapping programs developed in Australia, South Africa, Great Britain, and Europe, several botanists in Utah started the process of mapping herbarium collections using the 36 square mile township and range (approximately 10 km^2) grid system that is prevalent in the west. The development of an electronic atlas (Ramsey et al., in press) is an outgrowth of the Atlas of Vascular Plants of Utah project (Albee et al. 1988), and the development of an integrated GIS package for environmental management and conservation systems is an outgrowth of floristic inventories and studies of rare plants in Utah (Shultz & Hysell 1996; Sharik et al. 1998).

The GIS system we developed in Utah contains 1) historical records from herbarium specimens; 2) reports of distributions from published literature (monographs & technical reports); 3) species presence and abundance data from plots established to monitor changes in vegetation; 4) biophysical data provided by GIS and 5) on-site measurements of study plots established as part of a model to predict the distribution of rare species. Some of the rationale for our data structure and methodology is introduced in the following paragraphs.

Background

Our first study area was the 24,000 acre Camp W.G. Williams Utah National Guard training site in north central Utah. As part of a comprehensive botanical inventory, we developed a vouchered database and attribute file (Shultz et al. 1994) that provided key elements for coding environmental data. The botanical component of the GIS developed as a part of this project is a core element in analyzing ecological monitoring sites (Long et al. 1996). Later, we expanded studies into the Great Basin desert of western Utah, an 11,000,000 acre area included within the Hill Air Force Base Military Operation Area, in order to evaluate the utility of geographic information systems in predicting the occurrence of rare plants (Sharik et al. 1995). The data from these studies exist in a format that can be incorporated into state and national systems of information. The digital plant atlas provides data for all of Utah and is part of the geographic information systems developed for our two study areas.

GIS System

We started our project in 1993 using version 2.1 of Arc View. We are currently using Arc View 3.1 (Environmental Systems Research Institute, Inc. 1998). Much of the system's strength lies in its expandable and flexible data structure. We created and maintain files in a number of different programs, including Microsoft EXCEL, FoxPro. The Arc View program can import any file saved with a *.dbf extension as a table. Data tables can then be joined by using a common field. Shape files are treated as separate layer in the system. By using a common field, Arc View can join or link the separate data tables for spatial analysis. We found it convenient to create a genus/species (GENSPEC) code for each taxon in the file (expanding the code by letters or numbers when delimiting infraspecific taxa) and used numbers to identify each of the study sites. Geo-referenced bibliographic treatments and boundaries of populations of rare plants were digitized as polygons and treated as shape files. Since each record is spatially referenced in some way (either by points, grids, or polygons), all data in the system can be displayed on a map.

To date, the tables we are maintaining are 1) location grids from the Atlas of Utah Plants; 2) herbarium vouchers of all species of conservation concern; 3) location records reported in various publications or technical reports such as monographs or reports by Natural Heritage botanists; (4) new voucher collections for all plants collected in our inventories, and (5) observations of plants at specified survey sites. We assigned different symbols or colors to each of the tables, allowing us to identify the source of a report when viewing the mapped results of a query.

Development of the digital atlas

Maps from the Atlas of Vascular Plants of Utah (Albee et al. 1988) form the core of the working geographic information system for plants of Utah. Species maps were produced by plotting the distribution of approximately 400,000 Utah collections from the three major herbarium in the state: the Intermountain Herbarium of Utah State University, the Garrett Herbarium of the University of Utah, and the Brigham Young University Herbarium. Distributions were mapped by hand on a relief map at a scale of about 1:6,000,000. The results show approximately 73,000 different collecting locations for ca. 2,600 plant species. No specimen database was created with the project, but the original maps provide color-codes for the herbarium in which the record was found. Collection localities were then digitized on a uniform-area base map and integrated into our working GIS. Maps are available on an internet site (for web address, see citation with Ramsey et al.). While this referencing system has its faults (i.e., not referenced by individual specimen), the authors of the Atlas were able to spend their time examining specimens and making decisions about species circumscriptions rather than editing data files and supervising the development of a 400,000-record database.

New Voucher Records

As a part of the floristic surveys, we developed detailed voucher specimen files in simple spreadsheet format, providing specimen and locality data used to analyze species distributions. For our voucher records, we used fields used in the type specimens databases for the Gray Herbarium of Harvard and the New York Botanical Garden as well as additional fields for site data. Extracting the basic information, we generated labels for voucher specimens and produced various report files using report and label programs in the spread sheet or database program. Because we were not collecting records for the purpose of herbarium management, we did not use a program designed for management of systematic collections. However, by using standard fields of specimen information, we are creating databases that can be imported into any of a number of systems. In addition, we are providing detailed ecological information and vouchered references for distributions reported in the geographic information system.

Recent collection records usually provide reliable reports of geographic locations. For all of our collections, we used satellite-tracking technology in the form of Global Positioning Systems (GPS). Depending on our needs and whether or not we used a base station to make differential corrections, the GPS units allowed us to obtain geographic readings accurate to 2–20 meters. Geographic positions for new vouchers were recorded as latitude and longitude; township, range, and section; or u.t.m. (Universal Transverse Mercator) units. Latitude and longitude is a convenient unit of measure for Arc View, but any of these units can be translated into other systems or incorporated within a uniform grid system of any size. We found we could transfer files between different database programs for generating labels and reports. For analyses of spatial patterns and correlation of species distributions with environmental layers, we relied on the geographic information system.

Special data sets

For groups of special concern such as narrow endemics and invasive weeds, we developed separate data sets. These include specimen data and reports from the literature as well as field observations. In the specimen data set, we used the standard fields for herbarium collections (using the same system as that used for new vouchers) and a database for selected collections. Reports from published and unpublished literature were geo-referenced either by providing geographic coordinates or digitizing the boundary of an area onto a base map. These records were treated either as grids or polygons and cross-referenced to the appropriate bibliographic citation. Data collected at study sites include environmental data and observations of associated species. Study sites are numbered and geo-referenced either as polygons or points. Observations at study sites include lists of species, all of which are cross-referenced to geographic locations.

Organization and Query of Data Files

Query building tools that are part of the Arc View program allow us to find the locations of plants by any of the attributes provided in our table of authority and attribute files, such as taxonomic group (eg. family or genus), growth form (eg. tree, shrub, annual herb, etc.), habitat type, conservation status, flowering time, etc. In addition, we can create tables showing how species distributions correlate with other layers coded within the GIS, such as geological parent material or soil type. While some aspects of our system were programmed specifically for our project, much of our work was accomplished entirely with existing software.

The kinds of queries we can reasonably build from our files continue to expand as we develop attribute files and increase our number of study sites. Queries allow us to do simple yet otherwise time-consuming tasks, such as the following:

- develop species checklists by region or locality
- identify "hot spots" of endemism
- identify areas of high species diversity
- display data for voucher specimens
- show locations of species of conservation concern
- identify information gaps

A powerful use of the analytical tools is attained through ecological measurements coupled with tables of floristic data. In the studies based at Camp Williams, we coupled authority and attribute files with species-specific measurements from long-term monitoring plots established by ecologists (Long et al. 1996; Shultz & Hysell 1996). We were thus able to use the information system to analyze environmental data. Qualitative information about the species (i.e. weediness, growth form, and conservation status) coupled with quantitative measures in the field (i.e. frequency and size class) allowed us, for example, to track the invasions of exotic species after a disturbance, detect changes in species composition or growth form over time, and identify areas of conservation concern.

In one of our more interesting uses of the GIS, we tested its capacity to predict the distribution of rare plants in the western deserts of Utah. We were fortunate to have available one of the early state-level geographic information systems developed in collaboration with Federal agencies (Edwards et al. 1995) and we were able to use it to correlate distributions with a number of data layers. Then, by collecting additional biophysical data in the field (i.e. topographic position, landform, soil pH, soil texture, and associated species) at sites where species were present as well as absent, Aitken (1998) developed a hierarchical model to predict the potential distribution of four narrow endemics and established one of the first tests of a model to predict plant distributions. She compared the field models to other ones based on GIS layers and found that both GIS and field-based models performed well. The study demonstrates the value of location-specific data for plant distributions and is one that we hope will stimulate work with GIS-integrated floristic systems in other regions.

The floristic geographic information system allows us to create separate data layers from different information sources. For groups of special concern such as narrow endemics and invasive weeds, we developed a separate voucher database which uses core fields (i.e. collector, collection number, general locality, date of collection, location of herbarium collection, etc.) found in our new voucher files. Creating separate databases for historical data and current records made it easier for us to manage data files and code separate sources of information in the GIS. Historical records are treated as grid locations from the digital atlas of Utah plants herbarium, distribution statements from voucher records, or coordinates from the voucher files. In displaying all data layers on one map, we code each data table or shape file with different colors or symbols. Use of the "identify" button in Arc View allows a user to access the data layers associated with any point on a map, obtaining all the species reported from that position as well as the coordinates of the geographic position.

CONSIDERATIONS IN BUILDING GIS-BASED FLORISTIC SYSTEMS

Capturing basic data

Past experience by developers of herbarium databases provides some cautionary notes to developers of geo-referenced systems, but encouragement for efficient means of data transfer. In developing the PRECIS database for the *Flora of Southern Africa*, programmers found it useful to provide two formats for data output: (1) the "minis", consisting only of specimen identification, collector name, and locality (2) the "maxis", which reported all data stored for the specimen (Gibbs et al. 1984). Developers estimated that only about 10% of the specimens (from a total of 620,000 as of 1984) had grid references. The approach they took was to locate grid localities every time more than 20 specimens had been recorded in a particular location. In developing the digital plant atlas for Utah, we found a similar lack of grid references, and about 20% of the specimens that could not be mapped within a 10 km grid, no matter how hard we worked to determine a location.

The approach we used in Utah was to use a base map with topographic relief and map specimen distributions by hand, at times using the original field notes of collectors, but usually relying on simple judgment. For instance, finding a collection locality listed as "6 miles north of Logan along highway 89" and knowing the road turns in an easterly direction in that location, we could adequately map the specimen when we could not have relied on a computer to do so. Developers of the PRECIS system reported that the minimum data set was the one most often used.

Between the two extremes, i.e. creating a database for all herbarium collections or simply creating location records for herbarium specimens, is a solution that would take advantage of bar code technology and the relative ease of creating a simple data set. The creation of a wired (i.e. digital) link between a specimen, a grid location, and the name of a plant could be done with bar codes and scanners. Bar codes could be placed directly on specimens then digitally linked to a geographic position either by scanning an associated grid number or using a graphics table and pen-style plotter to mark a position on the map. Other information might be scanned from bar code tables (such as date of collection, coded by month and year), but decisions about what to add could be made at a later date. One advantage of such a system would be that information associated with the collections could be gathered as needed, or when feasible. A unique, electronically coded identifier can link the specimen to a geographic position on a map and provide a very efficient way to gather biodiversity data from herbarium collections. Specimens handled in this manner would have the advantage of having a grid-based spatial code that can be transported directly into a geographic information system.

Monographic Data

Monographic data provide another valuable source of information about species distributions, yet are usually overlooked because they exist in the form of a publication rather than individual specimens. As most botanists know, distribution data provided by a specialist are invaluable for identifying economically important resources and shaping conservation efforts (Prance 1994). The systematist who produces a monograph usually spends many years conducting field and herbarium studies of a particular group of organisms, in most instances providing the only fully documented source of information about a group of organisms and the only maps of species distributions. Ironically, one reason monographs have not been incorporated in data systems is that the records come from many sources and are not the propriety of any one collection, making data management difficult.

In the system we developed in Utah, a geo-coded reference to a bibliographic citation provided an efficient way to capture information from published maps. Distributions of species can be electronically coded by drawing an outline around the known distribution points, a mapping system such as the one employed to produce maps for the Flora of North America, or mapped as individual points. Using points or polygons, such a file could be imported into Arc View and indexed to an authority file. Capturing these spatially-referenced files in digital form should be a priority in the development of floristic databases.

Nomenclature and Attribute Files

The problem of changing systems of nomenclature and the need to develop an interactive classification system has been reviewed by Zhong et al. (1994) and is being addressed by other authors in this symposium. In North America, we are fortunate to have a synomymized checklist of North America plants developed by John Kartesz (1994) which incorporates the research of many contributors who are writing treatments for the ongoing Flora of North America project (Flora of North America Committee 1993a, 1993b, 1997). Nomenclature files provide an authoritative anchor for species names, and when maintained in a separate file within an integrated floristic system, have the advantage of being readily available for routine examination and cross-indexing to other data systems. Ideally, these files would be actively linked to bibliographic databases that incorporate all new records and combinations (such as the Kew and Gray Herbarium indices), but that task is beyond the scope of most herbaria. Instead, a simple nomenclature file consisting of the name of a taxon (genus, species, infraspecific name, etc.) and authority (i.e. author of basionyms and combinations), such as might be produced for a flora of an area, provides a relatively stable set of plant names. Synonyms can be compiled and maintained as a part of this 'names' file as long as a taxonomic authority is acknowledged for ascribing synonyms, such as the author of a flora, checklist, or monograph. Attribute files, providing information about individual species, can be developed as a part of a master list of names, then maintained as a part of the same file or segregated. The future holds promise for electronic access to attribute files developed in web-based systems such as the Biota of North America Project, the Flora of North America, the U.S.D.A. PLANTS database, and others (Shultz 1998a).

CONCLUDING REMARKS

Floristic botanists have massive amounts of data at their disposal. These data bring not only great advantages to the researcher, but enormously burdensome responsibilities for data management. We can and do become embroiled in issues of precision and scale and debates regarding the "best" system for organizing data at the same time that herbarium managers are overburdened with demands for information. The real cost of our continuing debate is the loss of perspective and productive time. Data are often managed as if species distributions are static, and we lose sight of the enormous potential in collection data to document changing environments and predict the responses of vegetation to that change. The work of systematists is essential to effective conservation planning and often crucial to the preservation of species and ecosystems. Data retrieval systems should simplify and amplify that work rather than complicate it.

Unleashing the power contained within our masses of data requires energy, and never has the energy been more limited. Systematists who know regional floras are rare, and are becoming more so with each passing year. Our discipline can ill afford to spend precious time cataloguing records or gathering information that already exists in other forms or can be compiled in other ways. Yet we know that a wealth of information is locked in herbarium specimens. We know that databases developed from collections provide details that cannot be analyzed because of highly variable information. Most will never be suitable for simple mapping programs, let alone detailed spatial analysis. What, then, is the solution?

In an address on the roles of natural history collections in society, Lane (1996) emphasized the importance of 'interconnectivity' among institutions and the need to develop shared technology in order to allow systematists to provide more robust answers to systematic, phylogenetic, and biogeographical questions. A new "KDI" (Knowledge and Distributed Intelligence) initiative from the National Science Foundation (NSF Bulletin, Feb. 1998) challenges us to develop computational systems that "focus on research and tools to discover, model, simulate, display, and understand complex systems or complicated phenomena; to control resources or deal with massive volumes of data in real time [and] to represent, predict, and design complex systems and understand their behaviors." The floristic data systems discussed in this paper were built originally from the most simplified data entry system possible: a position on a map linked by name only to a herbarium voucher. By using tools to unlock the data stored in natural history collections and the collective wisdom of floristic botanists, we can revolutionize our understanding of natural systems. The sooner we simplifying our systems for data collection, the sooner we can begin this important task.

ACKNOWLEDMENTS

Many of my colleagues in the College of Natural Resources of Utah State University contributed to the work reported in this paper. In particular, I am grateful to John L. Crane, James Long, David Roberts, and Terry Sharik in the Department of Forest Resources, and R. Douglas Ramsey in the Department of Geography and Earth Resources. Martha Aitken, Molly Hysell, and Tom van Niel had essential roles in developing the models used in the floristics lab, and in the Remote Sensing/GIS Laboratory of Utah State University, Frank Dougher, Kimberly Patraw van Niel, Pat Terletzky, and Thad Tilton developed the means to integrate the floristic databases into larger systems of analysis, including the Utah Gap Analysis program and environmental systems developed for the U.S. National Guard and Hill Air Force Base.

REFERENCES

AITKEN, M. 1998. A predictive model of potential habitat for four species of rare plants in the eastern Great Basin of Utah. Master of Science Thesis, Forest Ecology, Utah State University, Logan.

ALBEE, B.J., L.M. SHULTZ, and S. GOODRICH. 1988. Atlas of the vascular plants of Utah. Univ. Utah Mus. Nat. Hist. Publ. 270, Salt Lake City.

BEAMAN, R.S, A.G. THOMAS, and J.H. BEAMAN. 1998. A geographic information system for Mt. Kinabalu. In: J. Dransfield, M.J.E. Coode, and D.A. Simpson. Plant diversity in Malesia III. Proceedings of the third international Flora Malesiana Symposium, 1995. Royal Botanic Gardens, Kew. Pp. 425–430

EDWARDS, T.C., JR., C.G. HOMER, S.D. BASSETT, A. FALCONER, R.D. RAMSEY, D.W. WRIGHT. 1995. Utah gap analysis: An environmental information system. Utah Cooperative Fish and Wildlife Research Unit, Final Project Report 95-1, Logan.

FLORA OF NORTH AMERICA EDITORIAL COMMITTEE. 1993. Flora of North America north of Mexico. Volume 1: Introduction. Oxford University Press, NY. [*http://www.fna.org*]

FLORA OF NORTH AMERICA EDITORIAL COMMITTEE. 1997a. Flora of North America north of Mexico. Volume 2: Gymnosperms and Pteridophytes. Oxford University Press, New York, NY.

FLORA OF NORTH AMERICA EDITORIAL COMMITTEE. 1997b. Flora of North America north of Mexico. Volume 3: Magnoliidae and Hamamelidae. Oxford University Press, New York, NY.

GIBBS-RUSSELL, G.E. and P. CONSALVES. 1984. PRECIS–a curatorial and biogeographic system. In: R. Allkin and F.A. Bisby, eds. Databases in systematics. Academic Press, London. Pp. 137–153.

ENVIRONMENTAL SYSTEMS RESEARCH INSTITUTE, INC. ESRI Arc View© Version 3.1. 1999. Redlands, CA. [<http://www.esri.com/software/arcview/av31.html>]

KAREIVA, P. 1993. No shortcuts in new maps. Nature 365:292–293.

KARTESZ, J. 1994. Synonymized checklist of the vascular flora of the United States, Canada, and Greenland. Timber Press, Portland.

LANE, M.A. 1996. Roles of natural history collections. Ann. Missouri Bot. Gard. 83:536–545.

LONG, J.P., J.L. CRANE, E. OLSEN, T. VAN NIEL, K. PATRAW, and L. SHULTZ. 1996. Report on the national environmental database: A prototype from Camp W.G. Williams. Presentation to U.S. Army National Guard Bureau. Feb. 24, 1996. [Information about databases produced as CD-ROMs at the Remote Sensing/GIS Laboratory of Utah State University can be found at <http://www.nr.usu /ned/nedhome.html>]

NATIONAL SCIENCE FOUNDATION. 1998. NSF-Knowledge and distributed intelligence–proposal solicitation, p. 8. KN (Knowledge Networking), LIS (Learning & Intelligent Systems); NCC (New Computational Challenges). Report from the National Science Foundation: NSF9855.

PRANCE, G.T. 1994. The use of phytogeographic data for conservation planning. In: Forey, P.I., C.J. Humphries, and R.I. Vane-Wright, eds. Systematics and conservation evaluation. Syst. Assoc. Spec. Volume 50. Clarendon Press, Oxford. Pp. 145—163.

RAMSEY, R.D., L. SHULTZ, B.B. BANNER, and T. TILTON. A digital atlas of the vascular plants of Utah: Developing a GIS database from herbarium records, in review. [Maps provided at *http://www.nr.usu.edu/Geography-Department/utgeog/utvatlas/ut-vascatlas.html*]

REBELO, A.G. and S.A. TANSLEY. 1992. Where should nature reserves be located in the Cape Floristic Region, South Africa? Models for the spatial configuration of a reserve network aimed at maximizing the protection of floral diversity. Conserv. Biol. 6:243–252.

REBELO, A.G. 1994. Iterative selection procedures: Centres of endemism and optimal placement of reserves. Strelitzia 1:231–257.

Rhoads, A.F. and L. Thompson. 1992. Integrating herbarium data into a geographic information system: Requirements for spatial analysis. Taxon 42:43–49.

Scott, J.M., B. Csuti, J.D. Jacobi, and J.E. Estes. 1987. Species richness: A geographic approach to protecting future biological diversity. BioScience 37:782–788.

Scott, J.M., F. Davis, B. Csuti, R. Noss, B. Butterfield, C. Groves, H. Anderson, S. Caicco, F. D'Erchia, T.C. Edwards Jr., J. Ulliman, and R.G. Wright. 1993. Gap Analysis: A geographic approach to protection of biological diversity. Wildlife Monogr. 123:1–41.

Sharik, T.L. 1995–1998 Annual report on "population, distribution, and habitat study for threatened, endangered, and sensitive species of plants and animals within Hill Air Force Base Restricted Air Space and associated public lands, years 1–3. U. S. Air Force Environmental Division Technical Report, Ogden. [<http://www.nr.usu.edu/~hafb>]

Shultz, L.M., J.L. Crane, and M.T. Hysell. 1994. Botanical survey of a military reservation in central Utah. Amer. J. Bot. 81:200.

Shultz, L.M. and M.T. Hysell. 1996. Report of the floristic survey of Camp W.G. Williams Training Area, U.S. Army National Guard.. 76 p. + ArcView readable CD-ROM. [<http: www.nr.usu.edu/~tvn/campwil/campwil.html>]

Shultz, L.M., R.D. Ramsey, and N.R. Morin. 1998. "Floristics in North America: Tracking rare species electronically." In: C.I. Peng, ed. International symposium on the rare, threatened, and endangered floras of Asia and the Pacific Rim. Institute of Botany, Acad. Sinica Monogr. Ser. No. 16:259–273.

Shultz, L.M. 1998. The flora of the Colorado Plateaus: What do we know? In: L. Hill, ed. Learning from the land: Grand Staircase–Escalante National Monument Science Symposium Proceedings, Nov. 4–5 1997. Bureau of Land Management. UT/GI-98-006+1220. Salt Lake City. Pp. 203–209.

Zhong, Y., S. Jung, S. Pramanik, and J.H. Beaman. 1996. Data model and comparison and query methods for interacting classification for a taxonomic database. Taxon 45:16–28.

WEB SITE AND UNPUBLISHED DATA SETS FOR THE SOUTHEAST FLORA

NORTON G. MILLER and JORGE E. ARRIAGADA

Biological Survey
New York State Museum
Albany, NY 12230, U.S.A.

Department of Biological Sciences
St. Cloud State University
St. Cloud, MN 56301, U.S.A.

ABSTRACT

The amount of taxonomically useful botanical information available on the World Wide Web has increased greatly since the early 1990s, and responses by botanists to this important development have varied. We summarize our explorations of the Web for data pertinent to studies of the flora of the southeastern United States. Access to online library catalogs, bibliographies of journal articles, plant distribution atlases, authority files (e.g., authors of plant names, Gray Herbarium Index), lists of herbarium specimen holdings, and summaries of work in progress (often on "personal pages") are examples of what is available. For most of these, it is necessary to authenticate extracted information by visiting a library, confirming specimen identifications, and similar work. We evaluated examples of electronically accessible data sets to determine whether they are linked to specimens (and other primary sources) or derived and unlinked (secondary and tertiary sources). We concluded that while information derived from the Web may provide acceptable and prompt answers to many questions, using the Web uncritically and without verifying sources perpetuates errors and misinformation. A solution to this problem lies in the presentation and ability to evaluate metadata (statements about database scope and reliability) that should accompany all electronic sites. Seen in this light, botanical data assembled for access from the Internet will have a clearer link to library and herbarium collection resources that will always remain indispensable for research in floristics. Unpublished data sets about the flora of the Southeast present different problems–mainly ones of recognition by those needing information.

RESUMEN

Desde comienzos de 1990, ha habido un aumento espectacular en la cantidad de información botánica (taxonómicamente útil) disponible a través de Internet. Los botánicos han reaccionado diferentemente frente a la cantidad y calidad de esta información. En el presente trabajo, damos a conocer sucintamente lo que ha sido nuestra experiencia en la búsqueda de datos e información pertinente a la flora del sudeste de los Estados Unidos. Algunos ejemplos de información disponible en la red incluyen: catálogos de bibliotecas, bibliografías de artículos en revistas científicas, mapas de distribución de plantas, archivos de autoridades (i.e., autores de nombres de plantas, Gray Herbarium Index), listas de especímenes de herbarios, y resúmenes de trabajos recientes (a menudo ubicados en páginas personales). Sin embargo, en la mayoría de estos casos, es necesario verificar la información contenida en ellos mediante revisión de las citas bibliográficas originales, confirmación de la identificación de los especímenes citados, y de otras actividades similares. Con el propósito de determinar si los datos disponibles en la red están realmente basados en especímenes (fuentes primarias) o son datos derivados (fuentes secundarias y terciarias) hemos evaluado, a modo de ejemplo, algunos sitios. Es indiscutible que la información obtenida a través de la red proporciona respuestas rápidas y aceptables a muchas interrogantes. Sin embargo, el uso indiscriminado de esta información, sin verificar las fuentes, sólo perpetúa errores y desinformación. Una solución a este problema es la inclusión de metadata en todos los sitios electrónicos. Metadata incluye información sobre la cobertura de la base de datos, su fiabilidad, y su capacidad de autoevaluación. De esta manera la información obtenida a través de Internet tendrá una clara y directa conexión con los recursos bibliográficos y con colecciones florísticas. Los datos no publicados acerca de la flora del sudeste presentan variados problemas, especialmente relacionados con la aceptación de su validez por personas que necesitan la información para su uso inmediato.

Plant taxonomists, by and large, are making good use of the Internet and the World Wide Web (WWW) to disseminate information about the world's flora. We summarize here some of the botanical resources currently available on the Internet and in particular what may be useful for studies into the flora of the southeastern United States. This region has been the focus of a great deal of floristic research, including several regional syntheses, for example, the three editions of A.W. Chapman's *Flora of the Southern United States* (1860, 1889, 1897), J.K. Small's massive works (1903, 1913, 1933) and their rich content of information, several finished and in-progress floras of some of the Southeastern States or portions thereof, the *Vascular Flora of the Southeastern United States* started at the University of North Carolina at Chapel Hill, and the Generic Flora of the Southeastern United States begun by Carroll Wood in 1956 and continuing.

One of our purposes is to review how we and our Generic Flora coworkers use the WWW and

SIDA, Bot. Misc. 18: 83–96. 2000

other Internet resources in studies of the Southeast flora and to point out the benefits and what we perceive as some of the shortcomings of the Web for our research needs. The WWW is not a mature source of always-accurate information. The Web is in its infancy and is expected to grow and develop. New Web sites are established daily and others are lost. Furthermore, search engines, which locate and assemble information on the Web in response to specific queries, continue to be created, and those already available will change, fall into disuse, or be abandoned. Nevertheless, the Internet provides an unusual opportunity for rapid communication between persons and among research groups. Moreover, institutional or personal data-collections available now on the Internet as part of the WWW will continue to influence the conduct of floristic work in the Southeast and other regions, and especially as more Web sites are added. While we have found the WWW to be a useful mechanism to gather information about the flora of the Southeast, we have concerns about limitations of the Web and in some cases about the dependability of information retrieved from it. We explore these matters, focusing on information availability, scope, and reliability, and on how it can be best put to use in floristic studies.

Our expectations of floristic research are the following: (1) a new look at species, genera, and families represented in a well-defined area, including the adoption of accurate botanical nomenclature, (2) new assessments of information about species from primary and secondary sources and direct observation, (3) a thorough presentation of the geographic distribution of species that reflects contemporary field study and an evaluation of holdings of local and regional herbaria, and (4) workable identification aids, including keys and illustrations. We maintain that floras should have a large component of originality and also synthesize and present information confirmed by consensus and independent study.

To what degree can the Internet contribute to this process? Resources available now though the Internet include those specifically focused on the Southeast and others developed for larger geographic areas of which the Southeast is a part. We are critical of information and data derived from the Internet that is not anchored in either specimens or verifiable library resources, and we recommend ways that our concerns can be met by those developing Web sites.

THE INTERNET

The Internet is revolutionizing access to information. While libraries and librarians are often in the forefront of Internet access and services, in fact potentially anyone with a personal computer and a modem or some other interface has access to the Internet from practically anywhere, and with the right software can launch a Web site.

The Internet started from modest beginnings in the late 1960s, and about 1990 it began to grow significantly. In January 1999, the date of the most recent Internet Domain Survey, nearly 43 million hosts (typically defined as a computer with an Internet Protocol address) were identified throughout the world (Network Wizards 1999). The number of Internet hosts began to increase exponentially in the mid-1990s, and it has been predicted that in 2000 there will be 100 million of them (Rutkowski 1997a). The number of WWW hosts has also grown, and early in 1997 exceeded 400,000 (Rutkowski 1997b). In the last half of 1993, the Web was doubling in size about every three months, and in 1996 the doubling period had declined only slightly to less than six months (Gray 1998). Thus, significant growth in the Internet and its use seems a sure bet.

Components of the Internet that have had the greatest impact on our professional work include these: (1) e-mail, (2) online public-access library catalogs, (3) online bibliographic databases that are searchable using keywords, (4) authority files, (5) atlases of distribution records, (6) specimen databases, and (7) listservers. Web sites on the Internet contain examples of categories 4, 5, and 6. Some sites also have links to online catalogs (for example SUNY at Albany's home page links to the University Library [and to numerous other libraries], as does the New York State Education Department's home page to the New York State Library), and many provide e-mail to the site manager and others listed on the Web pages.

E-mail has greatly expanded the ease with which we and our botanical colleagues communicate. Its use has significantly shortened the response time to questions. In fact, we receive many more e-mail messages in a given period of time than letters delivered by the United States Postal Service. E-mail has affected our work in ways that are similar to the introduction of photocopying in the 1960s, namely faster and greater amounts of information capture. We now frequently transmit by e-mail partial or complete manuscript texts, allowing refinements to be made rapidly among co-authors.

Searching online public-access library catalogs for book and journal availability is quick and generally easy and can be accomplished from one's office computer. For example, although the New York State Library, a 19,000,000 book, document, map, and manuscript research collection, is located in the same building as our offices in the New York State Museum, we routinely call up EXCELSIOR, the NYS Library online catalog, when we need a book or journal and send a request for retrieval electronically. We also regularly enter HOLLIS, Harvard University's online catalog that includes holdings of the Harvard Botany Libraries, to ascertain book and journal availability, which helps us plan focused and productive visits to Cambridge, Massachusetts. The online catalog of the New York Botanical Garden (CATALPA) is also useful in the same way, as are those of Ohio State University (OSCAR), Cornell University, and many others.

We also use online bibliographic databases such as AGRICOLA (AGRICultural OnLine Access), which serves as a document locator and bibliographic system for the National Agricultural Library (United States Department of Agriculture); GeoRef, which abstracts selected records from the database of the American Geological Institute; AGRIS International, a comprehensive inventory of the agricultural literature worldwide; and Biological & Agricultural Index, which provides thorough, reliable indexing by librarians and subject specialists of 258 periodicals common to many libraries, including scientific journals. The New York State Library and many academic libraries have purchased subscriptions to search services, for example, Dialog@Carl plus RLIN and CitaDel, which provide access to 300+ bibliographic databases, abstracts, full-text articles, and directories, allowing an individual to customize a search and reach multiple databases simultaneously. At the New York State Library, these online bibliographic databases are available free of charge to registered library users, whereas access to other databases such as BIOSIS may incur a charge. Policies about database searches, including whether client- and/or librarian-directed searches are permitted, vary from library to library.

AGRICOLA is especially important in work toward a Generic Flora, because up to the early 1980s Carroll Wood and his co-workers at the Harvard Herbaria routinely examined copies of the *Bibliography of Agriculture* and typed card-file entries for all listed papers and books that were pertinent to the Generic Flora region (North Carolina, Tennessee, and Arkansas, south to Louisiana and Florida). This large file, now some 105,000 cards, is located at the Harvard Herbaria and is available to staff and visitors, although its existence is not widely known. It is used in the preparation of the annotated bibliographies that accompany the Generic Flora papers. Since new cards ceased being added to the file beginning about 1981, AGRICOLA back to 1980 is searched in connection with the preparation of the Generic Flora treatments.

Various authority files are available on the Web or via the Internet, including some that the Generic Flora Project has developed or has in preparation. An authority file provides standardized, consistent information to all users, thereby eliminating variations inadvertently introduced by authors that may follow different, but equally acceptable usage in another context. For example, a list of authors of plant names cited in all of the Generic Flora papers published so far is on line, and the abbreviations we use sometimes differ from those in other authority files, for example those of Brummitt and Powell (1992). A list of journals and standardized abbreviations of references cited in the published Generic Flora installments is also maintained, as are abbreviations for published floras that contain information pertinent to the plants of the Southeast. Our list of journals is more diversified than *Botanico-Periodicum-Huntianum* (Lawrence et al. 1968) and its *Supplement* (Bridson 1991), because our authors often cite the chemical and pharmaceutical literature. Journal abbreviations in the Generic Flora authority file grew from the list of botanical journals assembled by Lazella Schwarten and Harold Rickett (1958, 1961).

Apart from authority files that have been tailored for the Generic Flora Project, other authority files are available on the Internet. The Gray Herbarium Card Index, registers of type specimens held by various herbaria (FLAS, GH, NY, US), lists of plant collectors, and other examples are important in our work. Botanical authority files are often worldwide or hemispheric in scope, and thus have general applicability for any project and usually for all groups of plants. Atlases showing dot-map distributions (often one dot per county) and checklists of species known or reported from geographic areas in the Southeast and elsewhere are important data sources and provide guidelines for subsequent verification of plant occurrence. Examples include the Atlas of Florida Plants and those for Arkansas, South Carolina, and Tennessee. Plant specimen databases are less developed and perhaps more controversial than Internet resources in other categories, because of the high cost of preparation and maintenance. Relatively few comprehensive specimen databases are currently available for the Southeast or elsewhere, although Robert Haynes of the University of Alabama and his colleagues have been developing SERFIS, a specimen database about the holdings of 112 herbaria (over 6,000,000 botanical collections) in a region from the District of Columbia southward to Kentucky, Louisiana, and Florida. We admit to using such databases infrequently and rely instead upon traditional herbarium specimens borrowed from herbaria in the Southeast and nationally important herbaria that contain significant numbers of plants from this region, for example, the Gray Herbarium and that of the New York Botanical Garden.

INFORMATION ACCESS ON THE INTERNET

Specific sites on the Internet can be reached in a number of ways. Uniform (Universal) Resource Locator (URL) designations (addresses beginning with http (hypertext transfer protocol)) are for WWW sites; ftp (file transfer protocol) designations allow access and potential transfer of certain kinds of files; gopher addresses reach files that can be searched using keywords if WAIS search capability or another searching tool is installed at the specific site; and telnet designations that allow computer-to-computer linkages, including the ability to reach one's home computer from a remote site. In the Appendix, we provide short descriptions and addresses of databases and Web sites that we think are useful to students of the flora of the Southeastern States. The list surely is incomplete, and readers should also know that addresses may change and other sites may have come on line. For these reasons, it may be useful to employ an Internet search engine to help ferret out and classify new additions to the Web. Also, the Internet Directory for Botany is a useful guide to botanical resources on the Internet.

The Internet Directory for Botany is maintained and regularly updated at the Botanical Museum, Helsinki University, Finland, and the Herbarium, University of Regina, Saskatchewan, Canada, and mirrored elsewhere. The Internet Directory for Botany provides links to botanical Web sites in various countries in Europe, Asia, Africa, and North America that can be activated by mouse clicks. At present, the Directory is divided into 18 subject files, for example, arboreta and botanical gardens, botanical museums, herbaria, natural history museums, checklists and floras, and taxonomic databases, and botanical organizations. The Internet Directory for Botany provides a quick way to locate databases and Web sites pertaining to the Southeast and other geographic regions.

Another way to find needed information is to employ search engines. Search engines require one to provide keywords that structure the search process. Keyword choice and a successful search are often correlated. Whether one selects a string of keywords or uses Boolean logic ("and," "or," and "not" between keywords) is important. Reference librarians are usually well versed in search strategies, and with practice searches can be effectively performed by the novice. There are three types of search engines: general, multi-threaded, and subject specific. The differences between them are explained by Cohen (1997), who also provides much useful advice on the Internet's search capacity. To effectively employ a search engine its specific syntax must be followed. Multi-threaded engines are designed to search multiple search engines simultaneously. The multi-threaded approach is useful when subject-specific and general search engines produce unsatisfactory results. Search engines may be helpful in identifying new online sources of information. It should be stressed that the Web has grown so large that even the best search engines, for example, HotBot, in recent trials retrieved only about a third of the appropriate Web pages (Lawrence & Giles 1998).

The question of usefulness and quality of information retrieved from the Web has captured the interest of librarians and educators. For example, Knowlton (1997) in an article critical of the use of Internet resources in the preparation of college student research papers cites growing evidence that much information on the Web is not especially useful to scholars, although it may be helpful in planning a trip to Europe, deciding when to take the train between New York City and Washington, D.C., and so on. In fact, the Web has been likened to a bookstore in which Web sites (books) "exist because someone wants them there, not because any independent judge has determined them worthy of inclusion" (G.M. Santoro, as quoted in Knowlton 1997). Some sites that critically evaluate the Web and its quality are listed in the Appendix.

Not all Web sites contain suspect or uncritically assembled information. Those pertinent to the flora of the Southeast, for example, appear in all cases to have been carefully assembled. However, in keeping with the best traditions of floristic scholarship we suggest that botanical databases should provide a way for the user to trace the source of information back to its origin. In the case of dot maps, this would be the voucher specimen on which the electronic record rests. In the case of technical botanical information, the source is the beginning point, for example, where the protologue (and basionym) was published, to which useful bibliographic or secondary information can be appended when it facilitates a user going back to an original source or access to related sources. And, even if the bibliographic accuracy of an entry is verified, it still may be necessary to confirm observations, evaluate interpretations of competing explanations, or search for overlooked publications that establish priority.

In the writing of floras, information is sometimes copied from one source to another, without checking origin and accuracy. Facts are not always easy to verify when assembling hundreds of species descriptions or narrative discussions, but this should always be a goal. An example of a perpetuated error follows. Although it involves a series of books, the point we make nevertheless applies to information that is available electronically. The example was uncovered during the preparation of a treatment of the Smilacaceae for the Generic Flora of the Southeastern United States (Judd 1998).

Laurel-leaved greenbrier, or bamboo as it is often called in the South, *Smilax laurifolia* L., occurs on the Coastal Plain and sometimes the Piedmont, north to New Jersey, west to Texas, and up the Mississippi Embayment to Arkansas and Tennessee, and also in Caribbean Basin. Beginning with the first edition of A. W. Chapman's *Flora of the Southern United States* (1860) the flower was described as having one stigma and the fruits single seeded, whereas in the nearly contemporaneous second edition of Asa Gray's *Manual of Botany* (1856) the fruits were noted as mostly one seeded, but the stigma was not described. Subsequent editions of Chapman's flora (1889, 1897) repeat what is in the first. In edition five of Gray's *Manual* (1867) the stigma is given as solitary for the first time in the eight editions of this series, and the fruits are again noted as one seeded. This information is repeated or modified slightly through edition eight (Watson & Coulter 1889; Robinson & Fernald 1908; Fernald 1950), with one or more of the statements, ovary one celled (or one locular) and one seeded, ovary one locular, and ovules solitary. Stigma number appears as one or sometimes two in the first two editions of the Britton and Brown *Illustrated Flora* (1896, 1913; treatment of *Smilax* by Thomas Morong). However, in 1944, Coker more accurately pointed out that the ovary is two or three carpellate (but nearly always one locular at maturity by abortion) and one seeded, and the stigmas two or three, but soon deciduous. Edition three of the Britton and Brown *Flora* (Gleason 1952) does not mention stigma number nor do Gleason and Cronquist (1991). Similarly, Small (1933) is silent on this point, as are Radford et al. (1968).

Authority files, despite the name, may also contain errors, although ideally these should be few. Even so, it is wise also to retain some skepticism about the accuracy of information in authority files and to verify original sources. The widely consulted Gray Herbarium Card Index, a catalog of new names of vascular plants of the Western Hemisphere, has been converted from an index-card format to a digital one. This allows the Index to be reached on the Internet and questions to be asked of the database that earlier were not feasible, because a manual search of the Index would have been too time consuming. For example, it would have taken many days to go through the Index to prepare a list of new species of vascular plants published by the New York botanist Charles Horton Peck, whereas electronically, only a few seconds are required. However, mistakes or gaps in the Gray Index are occasionally discovered, and it is advisable, therefore, to check the original place of publication as well as other literature to verify or validate the accuracy of what is delivered by computer. Dr. K.N. Gandhi of the Gray Herbarium forwarded to us a selection of recent questions he received from Gray Index users. The questions pertain to the correct presentation and spelling of authors of plant names; missing entries; data not captured during digitizing; inconsistencies or ambiguities of data presentation; whether a certain journal run was examined for new names; typographical errors; and a variety of nomenclatural questions, including priority, later homonyms, and validly and effectively published names. The Gray Index accepts additions and modifications, and its quality (and that of similar authority files) is enhanced by such an option.

A well-prepared flora also rests on a foundation of carefully studied specimens. The Web and the Internet deliver at this time a modest amount of information about specimens (largely data on labels), but it is likely that current capabilities will greatly expand. However, the interpretation of herbarium specimens and label data can be subtle and a process not always suited to electronically transmitted data and images. Study of the actual specimen and the context in which it was gathered is often necessary, and this may involve library or other collateral sources. For example, the treatment of the Nartheciaceae in the southeastern United States (Zomlefer 1997) states that while *Aletris lutea* Small is reported from the *Gray's Manual* range (Fernald 1950), specimens in the Gray Herbarium show that this attribution was a mistake. The record was based instead on a supposed hybrid between *A. farinosa* L. and *A. aurea* Walter (but actually a color form of *A. farinosa*), and moreover, the Gray Herbarium held no specimens of *A. lutea* at the time Fernald was preparing the eighth edition of *Gray's Manual*. Our suggestion is that plant distribution data and maps made available electronically (or by other means) always be anchored on specimens, and that the vouchers be referenced in some way that is reachable by anyone needing to verify accuracy. The electronic distributional atlases available for Arkansas, Florida, South Carolina, and Tennessee can be traced back to the vouchering specimens.

Distribution maps have always been important in plant systematics as summaries of species occurrence that convey evolutionary and other historical implications. Because maps can now be efficiently produced by computer, combining geo-referenced locality data and plotting software, botanists should follow standards that allow access to the specimen data on which plant distributions are based. A printed distribution map often carries the assumption that it is accurate both geographically and botanically. For example, one of us (NGM) and his collaborators are preparing a flora of Rensselaer County, New York (Miller et al. 1996). We began by assembling a database of information about 4,209 specimens collected in Rensselaer County between 1835 and 1995 and preserved in the Herbarium of the New York State Museum. Locations for 70% of these vouchers were plotable on maps with varying degrees of accuracy (we used MapInfo software), but only 32% had precise coordinates. Many other specimens (38%) presented locality data with generalized, but still useful geographic data, but 30% could not be mapped at the scale we wanted, because the label reported only "Rensselaer County" or "east of Albany." Therefore, we established seven categories (Table 1) to explain the degree of precision that our maps show. Metadata of this sort are crucially important in the appropriate utilization of maps and similar graphic and data collections that are available on the Internet or by other means. Every presented data set should have fully explained parameters and limitations.

THE INTERNET AND FLORISTICS: PROMISE AND CONCERNS

We believe that computers and the Internet do and will continue to provide many opportunities to streamline the preparation and publication of floras. Floras of the future may be continuously

TABLE 1. Metadata Codes Used in the Mapping of Specimen Origin: Rensselaer County, New York, Flora Project. Percentages indicate proportions of 4209 specimens in each category. The vouchers are in the Herbarium of the New York State Museum (NYS).

O:	Coordinates appear on label. 6%
R:	Locations determined retrospectively by collector. 7%
W:	Location assigned by expert opinion. 18%
A:	Location estimated to be within a specific distance of the point listed in the database (e.g., within one mile). The town or city may not be known. 38%
T:	No point located, only town or city known. 25%
N:	No point located, town or city uncertain (e.g., Rensselaer County or east of Albany). 5%

updated by electronic protocols and have status-snapshots called up on demand. A small step in this direction has been taken for the bryophyte component of the Flora of North America Project. In operation as an experiment is an interactive system that allows the editors and authors to communicate with one and other as part of the manuscript submission and review process, both of which now take place over the Internet. The system is designed to allow centralized record keeping of all communications among participants and thereby to clarify the benefits and problems associated with this new form of communication. Individual taxonomic treatments, once they have been reviewed for content and edited, will be placed on the Web. This pre-print, electronic publication process allows anyone to use the preliminary treatments and to suggest refinements and contribute pertinent, omitted information, which can be transmitted to the author or editor for verification and potential incorporation.

As with any new technology, particularly one in which change is relentless, several concerns have arisen that in fundamental ways relate to traditions in plant systematics and floristics. Some have been discussed earlier in this paper and reflect the always-present need to check and verify original sources. Others pertain to mechanical aspects of new technology, such as the likelihood that an over reliance on the Web as a source of data will result in missed information and erroneous conclusions. Another concern relates to the preservation of databases that have been utilized in research and how a database or Web site is cited in a research paper. This is a complex issue, because databases may be changed regularly, even daily. If one extracts information one day, that information may not be preserved precisely as it was received and as it would be in a printed book. Some tentative steps have been taken to preserve periodic snapshots of the Web, but persons and institutions creating databases need to plan how each revision of a unique database can be preserved for retrieval (Gleick 1998). It has been suggested that successive editions of a database be backed up on a compact disk or preserved as a print out (Grube & Nimis 1997). Generic Flora authors have begun to cite Web sites as bibliographic entries, using URLs and the month and year that the database was read. In most cases, we do not know whether it will be possible to re-enter a database as it existed on the specific date it was used. This circumstance contrasts starkly with traditional library resources, which are more-or-less always at one's immediate disposal. It has been estimated that the average lifetime of a Web page is 75 days (*fide* B. Kahle, as cited by Gleick 1998). In contrast, our work sometimes requires us to read Linnaean or older ink on paper scarcely changed from its original condition.

Some problems we have encountered in reaching Web sites are these: unavailability (server down or extinct), location unknown (in some cases a possible address change), software incompatibility (thereby making a database difficult to download or retrieve information from); sites that have not been updated for several years; and authorized versus unauthorized access to databases and difficulties encountered in arranging access. Changes in or loss of Web site address over time is not trivial, in our experience. For example, of an original 88 sites listed in a draft of this paper dated April 1998, 32 (36%) could not be found (5) or had new addresses (27) in May 1999. All of these problems can have relatively simple solutions. For example, successive versions of databases should clearly present the last date of revision, as this may indicate information reliability. Moreover, and as mentioned earlier, when metadata are provided that explain content and characteristics of the database, the user can decide how best to meet whatever needs exist. In all this, however, the data and its manipulation and consolidation should not be seen as replacing the need to study plants. Someday, it may be possible to write a flora by picking and assembling information electronically, tapping into resources that others throughout the world have assembled. However, never leaving our offices to practice botany holds little appeal for us, even though we recognize that it is possible to comparison shop on the Internet from one's home and never visit a store! Insights from field study and first-hand observation and the application of new technology should go forward hand and hand.

APPENDIX

UNIVERSALLY USEFUL SITES

Aquatic, Wetland and Invasive Plant Database. Center for Aquatic Plants. Aquatic and Wetland Plant Information Retrieval System (APIRS) University of Florida, Institute of Food and Agricultural Sciences, USA

http://aquat1.ifas.ufl.edu/database.html

Access through telnet (128.227.242.241 or aquat1.ifas.ufl.edu)

Freshwater aquatic and wetland plants; incl. more than 40,000 research articles, books, and reports (taxonomy, ecology, physiology, utilization, control); references back to 1700s; 369 digitized color images of 192 aquatic plants and some line drawings, too.

Botanical Collector's Names from Index Herbariorum, Harvard University Herbaria, USA

http://herbaria.harvard.edu/Data/Collectors/collectors.html

Botany Databases at the Smithsonian Institution

http://www.nmnh.si.edu/botany/database.htm

Incl. Wood Collection, Lichen Databases, Index Nominum Genericorum.

United States National Herbarium Type Specimen Register at

http://160.111.81.6/gopher-menus/TypeSpecimenRegisterU.S.NationalHerbarium,US.html

Flora Europaea Database, Royal Botanic Garden, Edinburgh, UK

http://www.rbge.org.uk/forms/fe

Digital version of the *Flora Europaea*; incl. Atlas Florae Europaea, an overview of work toward mapping the distribution of European vascular plants.

Flora of North America (FNA)

http://www.fna.org

Information about the approximately 20,000 species of vascular plants and bryophytes of North America, north of Mexico; access to online library catalogs, related Web sites, online versions of the printed volumes; distribution maps (for authorized users only).

Geographic Names Information System, United States Geological Survey, Reston, Virginia, USA

http://www-nmd.usgs.gov/www/gnis/gnisform.html

Place names and their location.

Germplasm Resources Information Network (GRIN), United States Department of Agriculture Taxonomy Home Page

http://www.ars-grin.gov/npgs/tax/index.html

Taxonomic databases: Legumes; seeds of economically important plants (Association of Official Seed Analysts, International Seed Testing Association); noxious weeds; rare and endangered species. Distribution, common names, and literature citations for most of the vascular plants of North America.

Incl. Economic Plants Database (http://www.ars-grin.gov/npgs/tax/taxecon.html) and Ethnobotany Database (http://www.ars-grin.gov/~ngrlsb/).

The Global Plant Checklist of the International Organization for Plant Information

http://iopi.csu.edu.au/iopi

Incl. checklist of North America plants; still under development.

Gray Herbarium Card Index

http://herbaria.harvard.edu/Data/Gray/gray.html

List and bibliographic details of the names of American vascular plant species; names below this rank also included.

ILDIS International Legume Database & Information Service
http://biol.soton.ac.uk/ildis/LegumeWeb/
Synonyms, vernacular names, geographical records, uses, habitats, descriptors, literature references; incl. species from the southeastern United States; still experimental.
Also: http://www.ildis.org

Index Herbariorum
http://www.nybg.org/bsci/ih/ih.html
Only information for herbaria in the United States is regularly updated.

Index Nominum Genericorum, Department of Plant Biology, University of Maryland, College Park, USA
http://www.nmnh.si.edu/ing/
Validly published generic names of plants (including fungi); bibliographic citations and information about typification and nomenclatural status.

Index to Plant Chromosome Number, Missouri Botanical Garden, St. Louis, USA
http://mobot.mobot.org/Pick/Search/ipcn.html

Indices Nominum Supragenericorum, Department of Plant Biology, University of Maryland, College Park, USA
http://matrix.nal.usda.gov:8080/star/supragenericname.html
A joint project of the International Association for Plant Taxonomy and the Norton-Brown Herbarium, University of Maryland; lists valid and legitimate names of extant vascular plants above the rank of genus.
Alphabetical Listing by Genera of Validly Published Suprageneric Names
http://www.inform.umd.edu/PBIO/fam/inspvindex.html

International Code of Botanical Nomenclature. Tokyo edition (1994)
http://www.bgbm.fu-berlin.de/iapt/nomenclature/code/tokyo-e/

The New York Botanical Garden, Bronx, New York, USA
http://www.nybg.org
Index to American Botanical Literature at http://www.nybg.org/bsci/iabl.html
Specimen Catalog (incl. information about species of the southeastern United States); vascular plant types, bryophytes, macrofungi, and North American gymnosperms http://websun.nybg.org/bsci/hcol

The North American Flora
Plant Systematics and Evolution Links
http://www.csdl.tamu.edu/FLORA/tfp/tfplinks.html

PLANTS Database of the National Plant Data Center, United States Department of Agriculture, Natural Resources Conservation Service Ecological Sciences Division, Baton Rouge, Louisiana, USA
http://plants.usda.gov/plants/plnmenu.html
Checklists of the vascular and nonvascular plants of the United States. Lists of threatened and endangered species, noxious weeds, and wetland indicator status, environmental data, ecological site characterization, and native plant communities.
Also machine generated maps showing the distribution of plants of the United States, one dot per state at http://plants.usda.gov/plantssfr_scilist.cgi/

Royal Botanic Gardens, Kew, UK
http://www.rbgkew.org.uk/web.dbs/webdbsintro.html
Data from *Vascular Plant Families and Genera* (Brummitt 1992) and *Authors of Plant Names* (Brummitt & Powell 1992).

A Synonymized Checklist of the Vascular Flora of the United States, Puerto Rico, and the Virgin Islands
http://www.csdl.tamu.edu/FLORA/b98/check98.htm

TROPICOS, Missouri Botanical Garden, St. Louis, Missouri, USA
Nomenclatural database and associated authority files for vascular plants and bryophytes worldwide; a recently announced feature is queries of the bibliography for vascular plants (54,000 items) by author's last name or keyword at http://mobot.mobot.org/Pick/Search/pick.html

SOURCES SPECIFICALLY FOR THE FLORA OF THE SOUTHEASTERN UNITED STATES

ATLASES—

Arkansas Biodiversity, University of Arkansas Herbarium-BONAP

http://www.csdl.tamu.edu/FLORA/arkansas

An experimental, collaborative enterprise among faculty and students of University of Arkansas Herbarium, the Biota of North America Program, and Texas A&M Bioinformatics Working Group; information taken from *An Atlas and Annotated List of the Vascular Plants of Arkansas* (Smith 1988); incl. synonymy, citation of voucher specimens, common names, chromosome numbers, and distribution by county. The holdings of the University of Arkansas at Fayetteville Herbarium (UARK) contain voucher specimens for distribution map records.

Atlas of Florida Vascular Plants, Institute for Systematic Botany, University of South Florida.

http://www.usf.edu/~isb/projects/atlas/atlas.html

County distribution maps for 4,000+ vascular plant species; serves as an accessory to regional floras (Clewell 1985, Long and Lakela 1971; Wunderlin 1982); records of each taxon in all counties for which a verified herbarium voucher specimen exists, or in a few instances from authoritative literature sources. Records from University of Florida (FLAS), Florida State University (FSU), Fairchild Tropical Gardens (FTG), and University of South Florida (USF), also herbaria of the New York Botanical Garden (NY), Harvard University Herbaria (A, GH), and University of North Carolina, Chapel Hill (NCU); new records added regularly.

Atlas of Tennessee Vascular Plants

http://www.bio.utk.edu/botany/herbarium/vascular/atlas.html

Based on *Atlas of Tennessee Vascular Plants*, Vol. 1 (ferns, gymnosperms, monocots), 1993 by Chester, Wofford, Kral, Deselm and Evans and on *Atlas of Tennessee Vascular Plants*, Vol. 2 (dicots) , 1997 by Chester, Wofford, and Kral.

Checklist of Tennessee Vascular Plants

http://www.bio.utk.edu/botany/herbarium/vascular/chcklist.html

Based on Checklist of the Vascular Plants of Tennessee 1993 by Wofford and Kral.

Louisiana Grasses, Distribution and Diversity Maps, Herbarium, Department of Biology, Northeast Louisiana State University, USA, and Texas A&M University Bioinformatics Working Group

http://www.csdl.tamu.edu/FLORA/lagrasses/lahome1.htm

Creates distribution maps for species and infraspecific taxa; all taxa linked to corresponding distribution maps, and these can be examined by selecting a species epithet. Data files provides by NLU are derived from studies of the Louisiana Flora (Gandhi & Thomas 1989).

South Carolina Plant Atlas, Department of Biological Sciences, Clemson University, USA

http://cricket.biol.sc.edu/herb/

Approximately 3000 vascular plant species are listed, with a county dot map for each.

The atlas is based on Radford, Ahles, & Bell (1968) and follows Kartesz (1994); all distributional information based on at least one verified herbarium specimen.

DATABASES, CHECKLISTS, AND REFERENCES—

Checklist of Tennessee Vascular Plants - See entry under Atlas of Tennessee Vascular Plants

Field Guide: Dominant Flora of the Ground Cover in the Pine Forests of the Apalachicola National Forest, Florida, USA

http://bio.fsu.edu/htmls/overview.html

Checklist from Clewell (1985), Godfrey (1988), and Taylor (1992).

Florida Biotic Information Consortium, USA

http://www.flmnh.ufl.edu/fbic/

A group of scientists and librarians committed to identifying and making available information about Florida's biota.

Generic Flora of the Southeastern United States Project

A series of original taxonomic investigations of genera belonging to the 193 families of native and naturalized seed plants of an area including North and South Carolina, Georgia, Florida, Tennessee, Alabama, Mississippi, Arkansas, and Louisiana.

Files are available (for downloading) at the NYS Library anonymous ftp site (ftp://unix2.nysed.gov/pub/state.museum) in WordPerfect 5.1 format:

Read Me First (general background information)

Authors-Titles of Papers (published)
Chronological List of Papers (published)
Generic Flora of the Southeastern U.S. Authors of Plant Names
Generic Flora of the Southeastern U.S. Journal Abbreviations
Index - Families Published
Index - Genera Treated & Illustrated

Literature useful to the study of Florida plants, University of Florida Herbarium, USA
http://www.flmnh.ufl.edu/natsci/herbarium/bib
A selection of basic references related to the study of native, naturalized, and cultivated plants of Florida; incl. ethnobotanic and poisonous plant bibliography.

Southeastern Regional Floristics Information System (SERFIS), University of Alabama.
http:serfis.by.ua.edu/serfis.html
Database (telnet://serfis.by.ua.edu); incl. 100,000 records, 60,000 taxonomic and 120,000 geographic names; botanical collection data from these herbaria: University of Alabama (UNA), University of Florida (FLAS), University of North Carolina (NCU), and University of Tennessee (TENN).

Vascular Plants of North Central Florida, University of Florida Herbarium
http://www.fau.edu/divdept/biology/herbarium.htm
A checklist from Vascular Plants of Central Florida (Griffin 1978).

HERBARIA—

Florida Atlantic University Herbarium, Boca Raton, Florida, USA
http://www.fau.edu/divdept/biology/herbarium.htm
Flora of the southeastern Florida; incl. collections of *Ipomea* and a database of Convolvulaceae.

Herbarium of the University of North Carolina, Chapel Hill, North Carolina, USA
http://sunsite.unc.edu/unc-biology/herbarium/
Incl. image gallery of plants of the southeastern United States; under construction, to hold a Type Specimen Registry, database of scientific names, authorities, locations, label information, images of specimen labels, and annotations.

Tulane University Herbarium, Tulane University, New Orleans, Louisiana, USA
http://www.tulane.edu/~eeob/Facilities/Herbarium/index.html
Incl. list of type collections (148 vascular plants).

University of Arkansas Herbarium, Fayetteville, Arkansas, USA
http://www.csdl.tamu.edu/FLORA/arkansas/arkhome1.htm
Vascular plants and bryophytes of Arkansas and vicinity; Asteraceae of North America.

University of Central Arkansas Herbarium (UCAC) Conway, Arkansas, USA
http://www2.uca.edu/biology/bio_hp/herbarium/ucaherb.htm
Emphasis on the Arkansas flora; only a list of holdings is on line, but a searchable database is under development.

University of Florida Herbarium, Gainesville, Florida, USA
http://www.flmnh.ufl.edu/natsci/herbarium
Incl. bibliography of literature useful to the study of Florida plants; vascular plants of North and Central Florida. Type Specimen Catalog in development, currently includes 149 type specimens (http://www.flmnh.ufl.edu/natsci/herbarium/types).

University of South Florida, Institute for Systematic Botany and Herbarium, Tampa, Florida, USA
http://www.usf.edu/~isb
Checklist and bibliography of systematic literature of Florida vascular plants.

University of Tennessee
http://www.bio.utk.edu/botany/herbarium/index.html

BOTANICAL GARDENS AND ARBORETA—

Fairchild Tropical Garden, Coral Gables, Florida, USA
http://www.ftg.org
Collections of endangered, threatened, and rare species of South Florida; also cultivated and native plants of Florida.

Marie Selby Botanical Gardens, Sarasota, Florida, USA
http://www.selby.org
Type holdings, excluding Orchidaceae.

The State Botanical Garden of Georgia, Athens, Georgia, USA
http://www.uga.edu/~botgarden/
Native plants (woodland, and rare, threatened, and endangered species).

NATIVE, RARE AND ENDANGERED, AND POISONOUS PLANTS—

Florida Endangered and Threatened Plants
http://www.floridaplants.com/endangered.htm
Checklist of 54 endangered plants.

Poisonous Plants of North Carolina, North Carolina State University, Raleigh, North Carolina, USA
http://www.ces.ncsu.edu/depts/hort/consumer/poison/poison.htm
Links to Cornell University Poisonous Plants Page and to Medical and Poisonous Plants Database.

United States Fish and Wildlife Service Division of Endangered Species
http://www.fws.gov/r9endspp/endspp.html
Endangered and threatened species of the southeastern United States (The Red Book).

MEMBERSHIP ORGANIZATIONS—

Florida Native Plant Society
http://www.flmnh.ufl.edu/fnps/fnps.htm

SEARCHING THE NET

INTERNET DIRECTORY FOR BOTANY—
Organized by categories, incl. links to listed sites. Most sites are annotated or provide evaluations of the sites listed; search engine: http://www.botany.net/IDB/

Subject category list
http://www.helsinki.fi/kmus/botmenu.html

Cartographic links for botanists
http://www.helsinki.fi/kmus/cartogr.html
Internet sites with online plant distribution maps and links to sites with computer-generated maps.

Checklists, floras, taxonomic databases, vegetation
http://www.helsinki.fi/kmus/botflor.html
Sites for authors, chromosome numbers, collectors, glossaries, names, and for North America.

Software
http://www.helsinki.fi/kmus/botsoft.html
Incl. CASSIA (Collection and Specimen System for Information and Analysis); DELTA (Descriptive Language for Taxonomy, for writing taxonomic descriptions; incl. PANKEY (key generation and identification programs); FLORIN (for data management [taxonomy, nomenclature, geographic distribution, herbarium and living collections, bibliographies]); PANDORA (database system for biodiversity research projects); PLATYPUS (files for taxonomic, geographic, ecological, bibliographic data); TRANSLAT (Botanical Latin Translator Program); and MEKA (Multiple-entry Key algorithm, for rapid identification of biological specimens), http://shanana.berkeley.edu/meka/index.html; incl. WOODY (key to wetland woody plants of North America), ANGIOFAM (key to angiosperm families of the world), TROPIFAM (key to the woody plants of the Neotropics).

SUBJECT DIRECTORIES—

Argus Clearinghouse: http://www.clearinghouse.net/
INFOMINE: http://lib-www.ucr.edu/; scholarly Internet resource collections; located at the Library, University of California at Irvine; resources arranged by topics, e.g., biological, agricultural, and medical sciences.
Libraries on the Web: USA-Academic: http://sunsite.Berkeley.EDU/Libweb/usa-acad.html
Lycos: http://point.lycos.com/categories
Magellan: http://www.mckinley.com/

The Scout Report: http://wwwscout.cs.wisc.edu/scout/report/
Weekly updates including new sites of interest to the academic and research community.
Yahoo!: http://www.yahoo.com

SEARCH ENGINES—

Directory of search engines for botany:
http://www.helsinki.fi/kmus/botfind.html

Examples of general search engines:
Alta Vista: http://www.altavista.com
Hotbot: http://www.hotbot.com/
Infoseek: http://www.infoseek.com/
Google: http://www.google.com/

Multi-threaded search engines. (Simultaneous searches of many databases with one search statement.)
Cyber411: http://www.cyber411.com
Profusion: http://www.profusion.com
MetaCrawler: http://www.go2net.com (uses: Alta Vista, Excite, Infoseek, Lycos, Webcrawler, Yahoo!)
Plant-Link: http://www.plantamerica.com/palink.htm (URL search engine for plant related Web sites available; uses Metacrawler.)

Subject-specific search engines. (Searches of database collections assembled by topics.)
The Internet Sleuth: http://www.isleuth.com/
GovBot: http://ciir2.cs.umass.edu/Govbot/

EVALUATION OF INFORMATION FOUND ON THE INTERNET—

Sites, including criteria, used by scholars to evaluate information found on the Internet:
http://refserver.lib.vt.edu/libinst/critTHINK.HTM
http://www.albany.edu/library/internet/evaluate.html
http://milton.mse.jhu.edu:8001/research/education/net.html
http://www.tiac.net/users/hopc/findqual.html
http://www2.widener.edu/Wolfgram-Memorial-Library/webeval.htm
http://thorplus.lib.purdue.edu/research/classes/gs175/3gs175/evaluation.html
http://www.library.ucla.edu/libraries/college/instruct/web/critical.htm

CITATION OF ELECTRONIC SOURCES—

MLA (Modern Language Association) **style**:
http://www.uvm.edu/~ncrane/estyles/mla.html; also http://www.columbia.edu/cu/cup/cgos/idx_basic.html

ACKNOWLEDGMENTS

We gratefully acknowledge information, advice, and insights provided by librarians Stephanie Barrett (New York State Library), David Lane (University of New Hampshire), Heather Miller (University at Albany, SUNY), and Judith Warnement (Botany Libraries, Harvard University). Walter Judd and Wendy Zomlefer (University of Florida), Hugh Wilson (Texas A&M University), and Carroll Wood, Jr. (Harvard University) helped in various ways, and in particular shared their experiences with traditional and electronic methods in floristic research. Violeta Hernández edited and improved the Resumen. Gray Herbarium Card Index Bibliographer, K. N. Gandhi (Harvard University) provided information about the Index and its use. Our colleague at the New York State Museum, Wilma Campbell, helped us greatly with her thorough knowledge the Internet, library protocols, and wordprocessing skills. We also thank reviewer Anthony Brach for helpful suggestions. This paper is published as Contribution 774 of the New York State Museum and Science Service.

REFERENCES

BRIDSON, G.D.R. 1991. Botanico-Periodicum-Huntianum/Supplementum. Pittsburgh. Hunt Institute for Botanical Documentation.

BRITTON, N.L. and A. BROWN, 1896. An illustrated flora of the northern United States, Canada and the British possessions. Vol. 1. Ophioglossaceae to Aizoaceae. Charles Scribner's Sons, New York.

________. 1913. An illustrated flora of the northern United States, Canada and the British possessions. Ed. 2. Vol. 1. Ophioglossaceae to Polygonaceae. Charles Scribner's Sons, New York.

BRUMMITT, R.K. 1992. Vascular plant families and genera. Royal Botanic Gardens, Kew.

BRUMMIT, R.K. and C.E. POWELL. 1992. Authors of plant names. Royal Botanic Gardens, Kew.

CHAPMAN, A.W. 1860. Flora of the southern United States. Ivison, Phinney, & Co., New York.

________. 1889. Flora of the southern United States. Ed. 2. Ivison, Blakeman, & Co., New York.

________. 1897. Flora of the southern United States. Ed. 3. American Book Company, New York.

CLEWELL, A.F. 1985. Guide to the vascular plants of the Florida Panhandle. Florida State University Press, Tallahassee.

COHEN, L.B. 1997. Librarians on the Internet: The search for quality begins. Choice 34 (Suppl.):5–17.

COKER, W.C. 1944. The woody smilaxes of the United States. J. Elisha Mitchell Sci. Soc. 60:27–69 + pls. 9–39.

FERNALD, M.L. 1950. Gray's manual of botany. Ed. 8. American Book Company, New York and other cities.

GANDHI, K.N. and R.D. THOMAS. 1989. Asteraceae of Louisiana. Sida, Bot. Misc. 4.

GLEASON, H. A. 1952. The new Britton and Brown illustrated flora of the northeastern United States and adjacent Canada. Vol. 1. The Pteridophyta, Gymnospermae and Monocotyledoneae. Hafner Press, New York.

GLEASON, H.A. and A. CRONQUIST. 1991. Manual of vascular plants of northeastern United States and adjacent Canada. Ed. 2. The New York Botanical Garden, Bronx, NY.

GLEICK, J. 1998. The digital attic: An archive of everything. New York Times Magazine, April 12, 1998, pp. 20, 22.

GODFREY, R.K. 1988. Trees, shrubs, and woody vines of northern Florida and adjacent Georgia and Alabama. University of Georgia Press, Athens and London.

GRAY, A. 1867. Manual of the botany of the northern United States. Ed. 5. Ivison, Blakeman, Taylor, & Co., New York and Chicago.

GRAY, M.K. 1998. Web growth summary. http://www.mit.edu/people/mkgray/net/printable/web-growth-summary.html [April 1998]

GRIFFIN, D.G. 1978. Local flora: Vascular plants of North Central Florida. Rev. ed. Florida Cooperative Extension Service, Institute of Food and Agricultural Sciences, University of Florida, Gainesville.

GRUBE, M. and P.L. NIMIS. 1997. Mediterranean lichens on-line. Taxon 46:487–493.

JUDD, W.S. 1998. The genera of Smilacaceae in the southeastern United States. Harvard Pap. Bot. 3:147–169.

KARTESZ, J.T. 1994. A synonymized checklist of the vascular flora of the United States, Canada, and Greenland. Ed. 2. Vol. 1., Checklist, vol. 2., Thesaurus. Timber Press, Portland, OR.

KNOWLTON, S.R. 1997. How students get lost in cyberspace. New York Times, Education Life, Section 4A, November 2, 1997, pp. 18, 21.

LAWRENCE, G.H.M., A.F.G. BUCHHEIM, G.S. DANIELS, and H. DOLEZAL. 1968. Botanico-Periodicum-Huntianum. Hunt Botanical Library, Pittsburgh.

LAWRENCE, S. and C.L. GILES. 1998. Searching the World Wide Web. Science 280:98–100.

LONG, R.W. and O. LAKELA. 1971. A flora of tropical Florida. University of Miami Press, Coral Gables.

MILLER, N. G., C. FLOOD, W. F. BRODERICK, and J.E. MATTOX. 1996. The Rensselaer County, New York, flora resources survey. Program and Abstracts New York Natural History Conference IV, New York State Museum. Pp. 21–22.

NETWORK WIZARDS. 1998. Internet domain survey, January 1998. http://www.nw.com/zone/WWW/report.html [May 1999]

RADFORD, A.E., H.E. AHLES, and C.R. BELL. 1968. Manual of the vascular flora of the Carolinas. University of North Carolina Press, Chapel Hill.

ROBINSON, B.L. and M.L. FERNALD. 1908. Gray's new manual of botany. Ed. 7. American Book Company, New York and other cities.

RUTKOWSKI, T. 1997a. Internet trends-Internet hosts-overall trend. http://www.genmagic.com/Internet/Trends/slide-4.html [April 1998]

________. 1997b. Internet trends-WWW-prefixed hosts. http://www.genmagic.com/Internet/Trends/slide-6.html [April 1998]

SCHWARTEN, L. and H.W. RICKETT. 1958. Abbreviations of titles of serials cited by botanists. Bull. Torrey Bot. Club 85:277–300.

________ and ________ 1961. Abbreviations of titles of serials cited by botanists. Supplement. Bull. Torrey Bot. Club 88:1–10.

SMALL, J.K. 1903. Flora of the southeastern United States. Published by the Author, New York.

________. 1913. Flora of the southeastern United States. Ed. 2. Published by the Author, New York.

________. 1933. Manual of the southeastern flora. The University of North Carolina Press, Chapel Hill.

SMITH, E.B. 1988. An Atlas and Annotated List of the Vascular Plants of Arkansas. Ed. 2. Published by the Author, Fayetteville.

TAYLOR, W.K. 1992. The guide to Florida wildflowers. Taylor Publishing, Dallas.

WATSON, S. and J.M. COULTER. 1889. Manual of the botany of the northern United States by Asa Gray. Ed. 6. American Book Company, New York and other cities.

WUNDERLIN, R.P. 1982. Guide to the vascular plants of central Florida. University Presses of Florida, Tampa and other cities.

ZOMLEFER, W.B. 1997. The genera of Nartheciaceae in the southeastern United States. Harvard Pap. Bot. 2:195–211.

INCORPORATING PHYLOGENETIC RESULTS INTO FLORISTIC TREATMENTS

ROGER W. SANDERS

Botanical Research Institute of Texas
509 Pecan Street
Fort Worth, TX 76102-4060, U.S.A.

WALTER S. JUDD

Department of Botany
University of Florida
220 Bartram Hall
P.O. Box 118526
Gainesville, FL 32611-8526, U.S.A.
and
University of Florida Herbarium
Florida Museum of Natural History
379 Dickinson Hall
P.O. Box 110575
Gainesville, FL 32611-0575, U.S.A.

ABSTRACT

Although the foremost purpose of Floras is to provide accurate identifications, an important secondary function is to fix taxon concepts and names in the minds of users. In so doing, Floras should be based on current systematic knowledge. In this regard, the following philosophical issues are raised: 1) the precept that taxon concepts used in a Flora should be based on the principles of phylogenetic/cladistic systematics because only monophyletic taxa have historical reality, leading to more predictive and useful classifications; 2) the criteria for accepting novel taxon circumscriptions; 3) the implications of the application of alternative systems of nomenclature; and 4) the increasing divergence and isolation of phylogenetic and floristic researchers. Recommendations are made for implementing the current state of our knowledge of phylogenetic relationships into Floras, especially: 1) the urgent need for the continued support and contribution of data to TreeBASE, the prototype of a widely available, easily accessible, and frequently updated database on published cladistic analyses; 2) careful planning and knowledgeable, scientifically bold leadership by editorial committees of Floras; and 3) collaboration of molecular and traditional systematists, with emphasis on the importance of molecular researchers making phylogenetic results accessible for practical use.

RESUMEN

Aunque el principal propósito de las floras es proporcionar identificaciones precisas, una función secundaria importante es fijar conceptos de taxon y nombres en las mentes de los usuarios. Así pues, las floras deben basarse en el sistema normal de conocimiento. Con respecto a ésto, surgen los siguientes temas: 1) el precepto que los conceptos de taxon usados en una flora deben basarse en los principios de la sistemática filogenética/cladística porque solo los taxa monofiléticos tienen realidad histórica, dando lugar a clasificaciones más predictivas y útiles; 2) los criterios para aceptar circumscripciones de taxa nuevos; 3) las implicaciones de la aplicación de sistemas de nomenclatura alternativos; y 4) la creciente divergencia y aislamiento de los investigadores filogenéticos y floristas. Se hacen recomendaciones para implementar el estado actual de nuestro conocimiento de las relaciones filogenéticas en las floras, especialmente: 1) la urgente necesidad de un soporte y contribución continuados de datos a la TreeBASE, el prototipo de una base de datos ampliamente disponible, fácilmente accesible, y puesta al día con frecuencia de análisis cladísticos publicados; 2) liderazgo planificado de modo cuidadoso y conocido, científicamente audaz de los comités editoriales de las floras; y 3) colaboración de sistemáticos moleculares y tradicionales, con énfasis en la importancia de que los investigadores moleculares hagan accesibles los resultados filogenéticos para usos prácticos.

Floras, whether regional, national, continental, or global, function primarily to provide identifications. This usually entails, minimally, the use of diagnostic keys, as well as lists of taxa and their generalized distributions. Diagnoses/descriptions and distribution maps are also often included and aid in identification.

Furthermore, we believe that Floras should serve at least four secondary functions, which may not be obvious. The first is that they should provide an entry into the systematic literature of the

SIDA, Bot. Misc. 18:97–112. 2000

included taxa. That is, the sources used in the floristic treatment of each group should be cited. This will allow users to determine the rationale used to recognize the species and higher taxa as presented in the Flora. The second function is closely related to the first: Floras should provide thumbnail summaries of the current state of our systematic knowledge, highlighting remaining taxonomic problems relating to the included taxa (as a guide to needed research and collections). While this may merely involve citing current systematic, ecological, and ethnobotanical research, authors are often able to give valuable insights to difficulties that they had in sorting out the diversity in the group for that particular treatment. Floras exemplifying these functions include the *Flora North America*, in progress (FNA Ed. Comm. 1993), the *Generic Flora of the Southeastern United States* (Wood 1958), the *Manual of the Vascular Plants of Texas* (Correll & Johnston 1971), the *Flowering Plants of Jamaica* (Adams 1972), and *Genera Palmarum* (Uhl & Dransfield 1986). The third function is that Floras, especially those continental or subcontinental in coverage, should serve as references for the editors of journals, the authors of regional Floras and popular guides, and other professionals who want to follow an up-to-date standard for the limits and names of taxa within the region of concern. We contend that the fourth secondary function, whether intended by the authors or not, is to fix the concepts of taxa, especially families and genera, in the minds of users. That is, students, lay enthusiasts, and non-specialist professionals ultimately learn what a taxon "is" by using a Flora or by being trained by someone who is using a Flora. If professional systematists are honest with themselves, they will admit that the way in which they think of most genera and families dates back to when they first learned them, unless they intentionally update their concepts by following the literature. For example, do they prefer *Phyla* over *Lippia, Chamaesyce* over *Euphorbia,* or the converse? If so, is it because they were convinced by the presentation of research results that came to their attention after they had learned these groups for the first time? Consideration of these last two functions should make it clear that Floras have a powerful influence on the way both professionals and lay persons conceptualize taxa.

This paper is concerned specifically with the wealth of phylogenetic data that have been obtained by cladistic methodology, employing both DNA and morphology (the latter defined in the broad sense to include micromorphology, ultrastructure, anatomy, development, and secondary chemistry). Are the results of these cladistic analyses pertinent to the writing of Floras? If so, how are they to be incorporated? What are the practical concerns for those writing or editing Floras? Both of us use cladistic methods in our research and think that such methods comprise the logically most consistent methodological framework in which to study systematic relationships. Indeed, we submit and defend, below, that a Flora should be as phylogenetically informative as the current state of our knowledge allows. However, even if the editor(s) and/or authors are not so convinced, still they must deal with the cladistic literature from a perspective of reviewing and evaluating the published conclusions.

Stevens (1998) recently expressed similar concerns, especially in the context of understanding endemism in Malesia. He primarily addressed the writers of Floras on the criteria of the delimitation and ranking of taxa on the basis of phylogenetic principles in comparison to traditional (including evolutionary) principles. Our purpose here is to bring to light, in the context of the functions of Floras, an array of issues not addressed by Stevens.

These issues can be viewed, first, from the perspective of a comparison of philosophical and implementation issues. A second perspective distinguishes systematic and taxonomic issues from nomenclatural ones, all of which must be clearly differentiated when, especially, the philosophical issues, are discussed. Thus, we limit systematic issues to those dealing with the discovery of diversity and analysis of relationships. Taxonomic issues deal with the practice of using systematic data to recognize and conceptualize taxa and to convert hypotheses of systematic relationships into classifications, i.e., deciding on the limits of species and grouping those into higher taxa. Nomenclatural issues deal with assigning names to those taxa. Thus, different researchers may agree totally on the methods to infer relationships, while disagreeing strongly on the taxa to be recognized based on those inferences. Likewise, they may or may not agree on the nomenclatural conventions needed to name those differently conceived taxa.

The main body of the paper covers four philosophical issues. 1) With regard to the criteria for recognizing taxa, we defend the view that taxon concepts be based on phylogenetic relationships, and we review the current alternate views. 2) Criteria for evaluating and accepting/rejecting any given phylogenetic hypothesis are briefly considered. 3) Recent ideas that are unfolding concerning nomenclatural systems are explained and compared at some length, as this is an area that may be least familiar to most systematists. 4) Finally, we note the isolation of floristic researchers from phylogenetic researchers. This issue takes relatively less space to discuss, but it is perhaps of the greatest importance and leads directly into the consideration of the implementation issues that derive from the philosophical ones. Thus, by making other botanists aware of these issues and providing them a window to the pertinent literature, we hope to effect changes in the way writers of Floras have access to phylogenetic data, conceptualize taxa, and organize Floras to present these phylogenetic results to the public.

PHILOSOPHICAL ISSUES

ISSUE 1. What taxon concepts should Floras fix in the mind of users?

We believe that the overriding consideration for the classifications presented in Floras should be scientific accuracy. Other authors have debated the merits of multiple-use, multiple-user classifications (see Brummitt 1996; Stuessy 1997; Stevens 1990, 1998; and references cited therein). However, it is clear to us that all other purported uses, and ultimately the reputation of our field, stand or fall on the basis of scientific accuracy of the classification used. Although novel classifications and taxon circumscriptions may be disliked by some users, this should not be a concern. Furthermore, scientific accuracy will maximize the lifespan of the Flora. When floristic writers fail to provide a classification supported by current research, the Flora is out of date before being published, especially if an out-of-date or "fossilized" system is chosen as the basis of the classification and organization of the Flora.

This, of course, begs the question of what qualifies a classification for scientific accuracy. We submit that to be maximally scientifically accurate, and hence also maximally useful and predictive, the taxon concepts used in a Flora should be constructed by taxonomic principles based on concepts and methods of cladistics, i.e., phylogenetic systematics (hence, phylogenetic taxonomy, as used by DeQueiroz & Gauthier 1990, 1992, 1994; Bryant 1996; and others). This amounts to the monophyly criterion, i.e., taxonomic changes are necessitated when (and only when) taxa are shown to be para- or polyphyletic. Taxa above species are to be monophyletic and correspond to clades. This is in contradistinction to the application of traditional taxonomic principles (i.e., traditional or Linnaean taxonomy, also as used by DeQueiroz & Gauthier) by which taxa are justified by gaps only (monophyly and paraphyly are of no concern) or of the principles of the more recent version of traditional taxonomy, evolutionary taxonomy, by which paraphyletic taxa are explicitly accepted when justified by gaps. Of course, arbitrary systems, e.g., alphabetical or utilitarian, are considered to be scientifically and pedagogically unjustifiable, as well.

To many, this bias reflects not only an abandonment of the tried-and-true methods that have served for over two hundred years, but also an acrimonious devotion to a paradigm of doubtful validity. The debate over the last thirty years has led a great many researchers to acknowledge the logical basis of cladistic methodology and to accept phylogenetic systematics. However, the debate over applying principles of phylogenetic taxonomy versus evolutionary taxonomy is by no means over and is still capable of generating heated arguments.

The logical basis for phylogenetic or cladistic taxonomy (also called cladistic classification [Crane & Kendric 1997, Stevens 1998 among others] or even cladonomy, to signify derision [Brummit 1997]) is quite simple. 1) Historical groups exist in nature through time. 2) Only monophyletic taxa can have historical reality. This is because as soon as a subtaxon is removed artificially from a historical group and given equal rank to the remainder of the historical group, the historical group looses its integrity and no longer exists. (Of course, this remainder, which lacks historical reality, constitutes a paraphyletic taxon.) The argument for the recognition of monophyletic groups can be based on characters, as well. Paraphyletic groups can be diagnosed only on the basis of plesiomorphic (ancestral) characters, i.e., by the lack of derived characters that diagnose the subtaxon.

The evolutionary taxonomic philosophy and arguments voiced by Cronquist (1987), Brummitt (1996, 1998), Sosef (1997), Brummitt and Sosef (1998), Stuessy (1990, 1997), and references therein, which conclude that paraphyletic groups are unavoidable and, indeed, scientifically valuable, are flawed. Counter-arguments have been well-articulated by Donoghue and Cantino (1988), Humphries and Chappill (1988), and Freudenstein (1998). (See also Welzen 1997, 1998; Schander 1998b; and Stevens 1998; and references cited therein.) A few points relating to these recent criticisms of the phylogenetic approach by the advocates of the evolutionary view are discussed below.

1.—The evolutionary taxonomic viewpoint misapplies the terms monophyletic and paraphyletic to species. Various species concepts bear on this misunderstanding and the reader is referred to the references cited within the above papers for details (see also Davis & Nixon 1992, DeQueiroz & Gauthier 1992). We note that, in general, species, as phylogenetic units, are fundamentally different from clades. Species cohere by reticulating relationships (either by gene flow or stasis of gene frequencies) of their constituent individuals and populations. Clades "cohere" by hierarchical relationships of their constituent species. Hierarchical concepts and rules cannot be applied to species, the units on which the hierarchy is built. Thus, a species is neither monophyletic nor non-monophyletic.

2.—It fails to understand that genera and higher taxa do not give rise to other genera and higher taxa, only species are ancestors of other species and higher taxa. When a small (more exclusive) clade is recognized as a monophyletic taxon within a larger (more inclusive) clade, its sister group either will be another clade, which should also be recognized as another monophyletic taxon, or else the sister group will be a species. Even if that species is the ancestral species, it follows from point 1 above, that the species is neither paraphyletic nor non-paraphyletic. Hence, paraphyletic (supraspecific) taxa are not inevitable.

3.—It confuses ranks and taxa. The rank is not the taxon, which is circumscribed to consist of a particular set of species. The taxon retains that circumscription regardless of the rank to which it is assigned (and the name it receives at that rank). Just because one desires to recognize different taxa, they do not have to be placed at the same rank.

4.—It fails to determine the level of universality of the characters observed. When a paraphyletic taxon is created by the recognition of a monophyletic segregate at the same rank, the decision is based, at least in part, on the occurrence of synapomorphies in the segregate and the alternate state (symplesiomorphy) of each of these characters in the remainder. Evolutionary taxonomy equates the alternate states at that rank, using the symplesiomorphies to diagnose the paraphyletic group. Actually, the alternate states should be treated as synapomorphies at higher levels of universality where they diagnose more inclusive clades. For example, the development of eusporangia is a synapomorphy of all embryophytes, not of eusporangiate ferns. Likewise, gymnospermy may appear to be the alternate state to angiospermy. Upon closer inspection, however, seed presentation in non-flowering seedplants is found to be diverse, and there is yet no consensus regarding which of these diverse structures gave rise to "seeds enclosed in a carpel." In the absence of such knowledge, the most proximal ancestral (and, hence, alternate) state leading to angiospermy is simply the presence of seeds (i.e., "megagametophyte persistent and enclosed in integument") which is a synapomorphy of all seedplants, not just "gymnosperms."

5.—It confuses cladograms and phylogenetic trees. In cladograms, the nodes are hypothetical ancestors. Such diagrams show only sister-group relationships, not ancestor-descendant relationships, which are very difficult to ascertain.

Thus, Brummitt's concern about the "telescoping effect" of all higher taxa back into the ancestral species is unnecessary. Recognition of internested monophyletic taxa is not only possible but also desirable, especially when taxa traditionally given the same rank are shown to be at very dissimilar levels in the phylogenetic hierarchy. **Paraphyly results from trying to show "ei-**

ther-or" character states by forcing internested taxa into the same rank, making them appear to be sister groups and, therefore, distorting the phylogenetic pattern.

AN EXAMPLE

Brummitt (1996) criticized our earlier paper (Judd et al. 1994), in which we concluded that the Cruciferae and Capparaceae should be combined. A defense of that recommendation serves to illustrate the consequences of these opposing positions. We preface this comparison by pointing out that Brummitt misunderstood the thrust of the paper when he stated (p. 380), "My first objection is not that it is inconvenient or disruptive to lose these well-known families, but that no indication is given of the resulting infrafamilial classifications." Yet we said (p. 1):

> Our intent here is to perform preliminary cladistic analyses of these family groups in order to explicitly test the proposition that the "tropical" member of each of these family groups is paraphyletic. Until phylogenetic relationships within each of these groups of families are clarified, it is not possible to address other significant macroevolutionary, historical, or taxonomic questions such as: ...4) Are the members of these family groups circumscribed in a non-arbitrary fashion?

Cladistic methods can be used to test specific hypotheses, as we did. While our methods were appropriate for testing that hypothesis, they were not appropriate for testing infrafamilial relationships, which were irrelevant to our defined task. Thus, we pointed to the need to realign family concepts. We were constrained by our data from discussing delimitation of taxa at lower ranks; to do so would have been irresponsible.

With respect to the Capparaceae/Cruciferae, our objective was to show that the Capparaceae, as traditionally conceived, are paraphyletic. This should be apparent, even to the thoughtful noncladist, because the Cruciferae (s. str.) comprise only a single major line of specialization with a limited range of variation; whereas, the Capparaceae comprise several lines of specialization with a broad range of variation. Brummitt's major objection, of course, was to the submerging the cruciferaceous clade into the capparaceous in order to correct the paraphyly of the Capparaceae (which under the Code must take the name Brassicaceae or Cruciferae). He asked, "...is it really necessary to obfuscate the non-botanist's concept of the Cruciferae by including in it the Caper, *Capparis spinosa* L., with a conspicuous mass of stamens in every flower? Who benefits from this?"

Our response is that all users of the classification benefit. We did not destroy the taxon, Cruciferae s. str. (it is still there to be conceptualized by anyone), but merely put it in its scientifically proper, historical relationship (as far as our data would allow; Figure 1; Table 1, 2). To use one of Brummitt's arguments (p.374) in reverse, if the Cruciferae s. str. comprised only two species, would he still insist that they be separated at the rank of family? From a slightly different perspective, Stevens (1998) came to our defense in stating:

> Group circumscription in evolutionary classification [equivalent to our use of evolutionary taxonomy] is subject to so many but at the same time such weak controls that usefulness and convenience speedily become ends in themselves. Let us never again hear of systematists "obfuscating" non-botanist's concepts of taxa like *Cruciferae* as if that were a valid argument against changing their circumscription.

Furthermore, it seems odd that, to make his point, Brummitt would specifically refer to the androecium. Why then does he allow *Cleome*, with its mere six stamens, to be included in the Capparaceae if a conspicuous mass of stamens is integral to the concept of the family? We suspect it is because the androecium of *Cleome* violates his (as well as the non-botanist's) concept of the Cruciferae, based perhaps on a too narrow sampling of species. Presumably he would say that the stamens of *Cleome* are long and showy, resembling more those of the Capparaceae. However, one should remember that the basal genera of Cruciferae s. str., such as *Warea*, have stamens that are longer, showier, and less tetradynamous than those typical of crucifers.

Finally, one can easily show that the Capparaceae lack historical reality; other than being an arbitrary taxonomic concept, they are a non-entity: This is because they simply lack any unique, defining characters (Table 1). Can Brummitt or anyone else give a simple, clear, unequivocal characterization of the Capparaceae without invoking a contrast to characters defining the crucifers? We cannot.

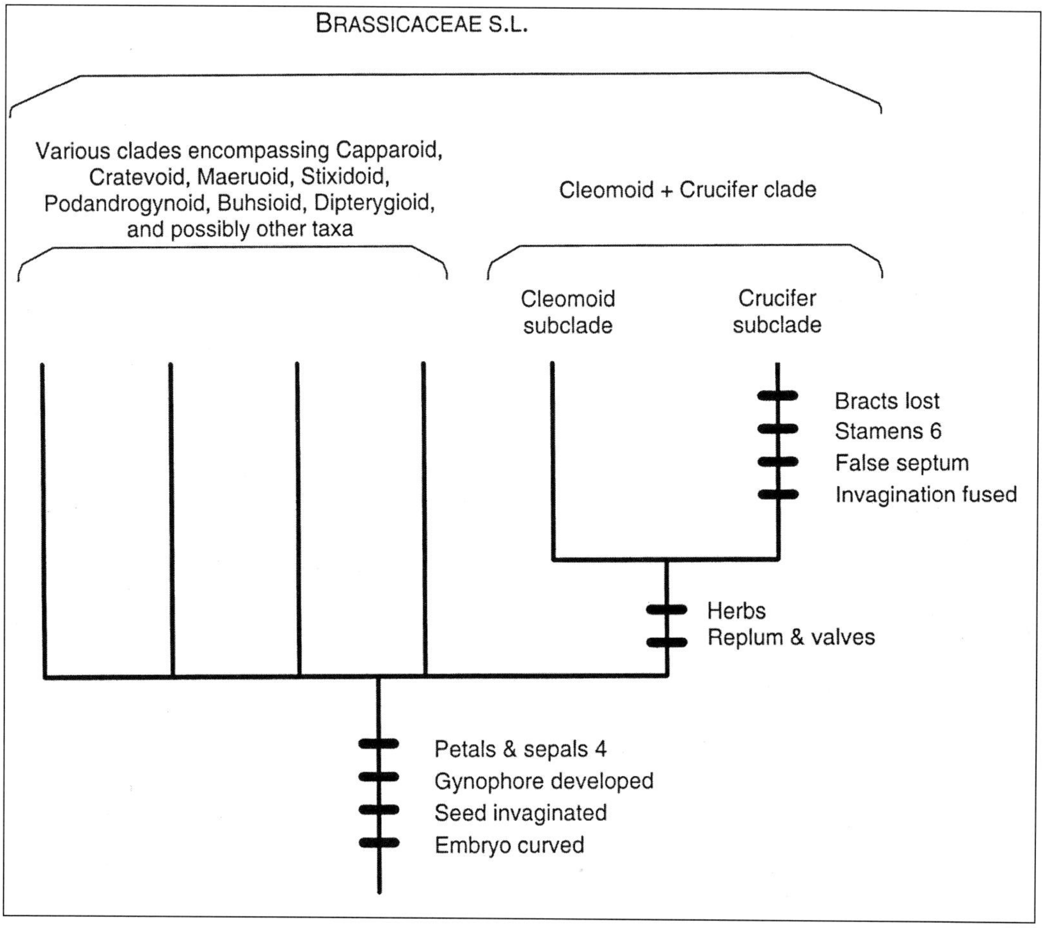

FIG. 1. Cladogram of Brassicaceae *s.l.* in support of Tables 1 and 2, emphasizing only those characters necessary to show the paraphyly of the Capparaceae (and hence, largely unresolved for those taxa) and the imbedded position of the Cruciferae *s.st.* Based primarily on Judd et al. (1994).

Thus, Brummitt's conclusion—that, in a cladistic classification [phylogenetic taxonomy, in our usage], taxa are defined principally on the putative phylogenetic position, whereas, in Linnaean classification, [traditional taxonomy, in our usage] taxa are recognized by characters—is a false comparison. Rather, phylogenetic position is determined from characters; not from the gaps within characters (as in traditional taxonomy), but from synapomorphies, i.e., the presence of character states in which the level of universality has been correctly interpreted (Table 2). Therefore, we must conclude that those who explicitly recognize paraphyletic taxa use characters inappropriately to discover taxa that do not exist!

ISSUE 2. What are the criteria for accepting a revised classification for use in a Flora?

Obviously, authors (ourselves included) of systematic research hope to have their conclusions accepted. However, it is the responsibility of the writers of Floras to decide whether or not to accept those classifications. We recommend that clades to be recognized as named taxa should have strong support, rather than being based only on preliminary evidence. There should be a minimum of two, and preferably several lines of corroborating evidence. One of these lines should be morphology (in the general sense) and, the other, molecular (DNA). If a cladistic analysis of morphology has not been published, then several genes should be sufficient (but see comments under implementation issues). Some measure of clade support, e.g., decay, bootstrap,

Table 1. Comparison of characters used to define the Cruciferae and Capparaceae.

Characters	Cruciferae s. st.	Capparaceae
Habit	herbaceous	variable, woody as in close outgroups but also herbaceous as in Cruciferae
Floral bracts	absent (i.e. lost)	present, as in close outgroups
Sepals and petals	4	4, as in Cruciferae
Stamens	6 (or fewer)	variable, numerous as in close outgroups but also 6 as in Cruciferae
Gynophore	variable, elongate in basal group(s), reduced or lost in typical groups	usually elongate as in basal Cruciferae, occasionally reduced or lost as in typical Cruciferae
Ovary/fruit locules	2, with false septum	1 without false septum, as in close outgroups
Fruit dehiscence structure	valves falling away from replum	variable, unspecialized as in close outgroups but also having valves falling away from replum as in Cruciferae
Seed wall	invaginated and fused	intermediate, invaginated as in Cruciferae, but adjacent sides not fused as in close outgroups
Embryo	curved	curved as in Cruciferae

TABLE 2. An indented, phylogenetic taxonomy of the Brassicaceae *s.l.* (combined Capparacee and Cruciferae) using apomorphies to show how historical, evolutionary relationships are better represented than in a traditional taxonomy of two coordinate families.

BRASSICACEAE S.L. (including Capparaceae)	
	Petals & sepals 4 Gynophore developed Seed invaginated Embryo curved
CLADE OF CLEOMOIDS + CRUCIFERAE S.ST. (among other major clades of family)	
	Herbs Fruit dehiscent by replum and valves
SUBCLADE OF CRUCIFERAE S. ST. (sister group of Cleomoid subclade)	
	Floral bracts lost Stamens reduced to 6 (as in some other lines of the more inclusive groups) Ovary/fruit bilocular by development of false septum (silique fruit) Seed invagination fused close

or jackknife analysis) should be available. The analyses based on different character sets should be congruent and compared, and/or the characters included in a single mixed-data analysis. Also, repeated studies using more complete sampling and independent evaluation of the characters would constitute stronger support than would a single analysis.

The problem here is how to define "strong support." Cantino et al. (1997) have suggested various operational criteria to quantify the level of support a clade receives. Wagner and Beitel (1992), Backlund and Bremer (1998), Stevens (1998; see also comments in Stevens 1997), and Judd et al. (1999) each suggest subsidiary criteria for whether well-supported clades should be recognized as taxa and at which relative ranks.

Authors of Floras should care enough about the users not to give a phylogenetically misleading classification, especially if there is strong support for a more accurate one. Likewise, if a revised classification lacks support, it should be not used in the Flora just because it is new.

A related concern is the effect on the stability of taxonomic concepts and names brought about by applying phylogenetic taxonomic principles. Under our current Code, phylogenetic classification may have disruptive effects as paraphyletic groups are divided into more numerous monophyletic taxa or are combined with their segregates. At the genus level, both kinds of changes necessitate transfers of species among genera. Recognition of all well-supported clades may inflate the number of ranks (although there may be either an increase or decrease of numbers of taxa at any particular rank). However, as more and more groups are studied, the well-supported clades are likely to persist and taxonomic stability in the long run will be greater than when traditional taxonomic methods are used. Indeed over the last 200 years, the taxonomic community has accommodated multitudinous changes in classifications and names based on *ad hoc* criteria, even whims. Thus, we conclude that taxonomic stability is a concern only when it does not conflict with scientific accuracy.

Issue 3. Proposed alternatives to the traditional (Linnaean) system of nomenclature.

For a number of years, systematists have been unhappy with certain aspects of traditional nomenclature that are philosophically rooted in the Linnaean hierarchy, which serves as the basis of traditional taxonomy. Recall that in the Linnaean system, a name is applied to the rank at which the taxon has been assigned and not directly to the taxon; at that rank, the name is permanently attached to an element of the taxon, the nomenclatural type, not directly to the taxon. Apart from their rank-based suffixes, taxa at subsidiary ranks often lack unique names. Finally, every species must be assigned to certain mandatory ranks. Strict adherence to the Linnaean system of nomenclature, as currently practiced, results in the following sorts of confusion and instability (Fig. 2):

- The choice of rank to which a taxon is assigned is arbitrary, based on a mix of tradition and preference.
- The name of a taxon changes with its rank assignment, even though the contents of the taxon do not change.
- A name at one rank can apply to two or more competing concepts of taxa.
- Names of higher ranks for monotypic taxa, such as Ginkgophyta and Ginkgopsida for *Ginkgo biloba,* are redundant and lack information.

Recently, concern has developed that strict adherence to the Linnaean system of nomenclature is in conflict with the goals of phylogenetic taxonomy. That is, the current codes of nomenclature favor the goals and perpetuation of traditional taxonomy over the goals and adoption of phylogenetic taxonomy. This conflict arises from two sources. First, traditional nomenclature is a logical outcome of the Linnaean hierarchy, which consists of abstractions called ranks; taxa are assigned to a limited number of fully internested ranks but, at any one rank, are mutually exclusive. In contrast, the phylogenetic hierarchy consists directly of taxa in multiple sister-group levels (i.e., nested sets of monophyletic groups). Second, in traditional nomenclature as noted above, a name is not unique to the taxon, whereas historical groups require unique, permanently attached names.

As a result, a continuum of responses has developed, centered around three schools of thought. 1) The most conservative school, consisting of both traditional and phylogenetic taxonomists, supports maintaining the Linnaean system as in the current Code. Some evolutionary taxonomists, such as Cronquist (1981) advocate type-based names at all higher ranks. 2) The moderate position is to modify the Code to relax the requirements that are most in conflict with phylogenetic goals, while preserving other valuable rules. 3) The most radical response is a proposal to overthrow the Linnaean system and replace it with a phylogenetic system of nomenclature. This new system currently exists only in principle as it lacks a written code. Our purpose here is not to espouse any particular view but to summarize the literature centering on this debate. However, we are opposed to following any system for which there is no written code, i.e., name changes should not be made in a *ad hoc* fashion.

Because the conflict in the two systems has been brought into sharpest focus by the advocates

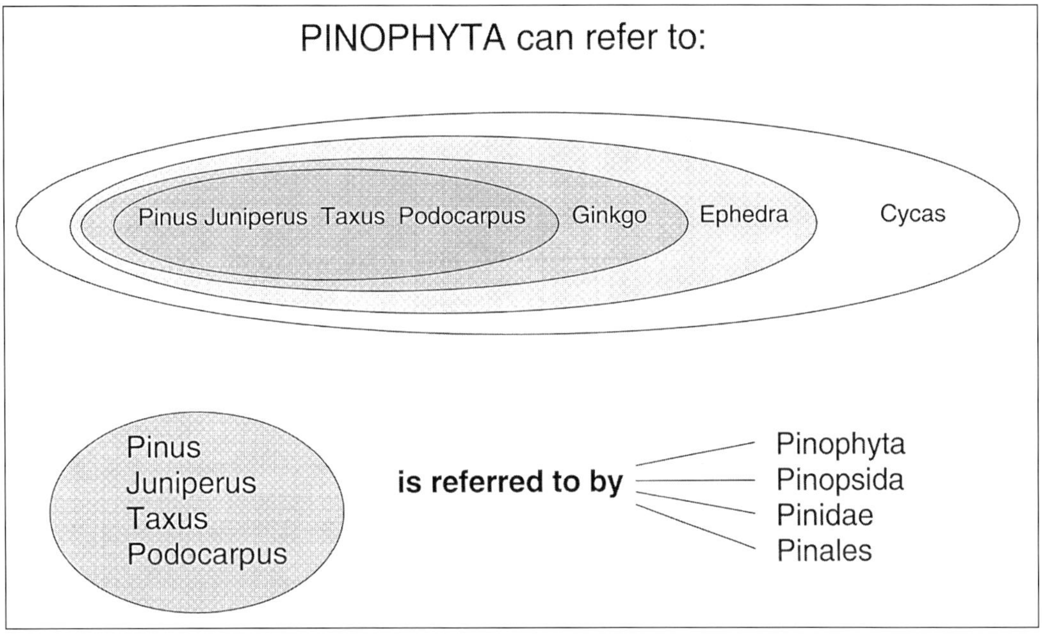

FIG. 2. Example of the major forms of confusion caused by type based names, regardless of whether taxa are based on methods of traditional (including evolutionary) or phylogenetic taxonomy.

of a phylogenetic system of nomenclature, we examine first the pros and cons of this school of thought as we see them. The basic principles have been articulated primarily by De Queiroz (1996, 1997) and by De Queiroz and Gauthier (1990, 1992; 1994; see also modifications and/or critiques by Bryant 1994, 1996; Moore 1998; Schander 1998a; Welzen 1998; and counter-arguments by Brummitt 1997; Lindén et al. 1997) as follows:

- No taxonomic ranks (mandatory or optional) are to be used. Only two catagories are recognized, species (non-hierarchical units) and clades (hierarchical units); the species and clades are conceptualized and named as taxa.
- Each taxon has a unique name; theoretically this applies to species, as well.
- Names of clades (not species) are to be defined by reference to ancestors, not by reference to a nomenclatural type of the taxon at a particular rank.

For example, the taxon, Brassicaceae, is defined under the ICBN as, "a taxon at rank of family in which *Brassica* is a member." If, however, it were defined as, "the taxon that includes the most recent common ancestor of *Brassica* and *Crateva* and all of its descendants," the name Brassicaceae would become unique and permanently attached to this particular groups of species. This would be true even if the rank were changed from that of family or if it were given no rank at all.

Several experiments have been published in which phylogenetically classified taxa were provided with names that were both formulated by the principles of phylogenetic nomenclature and were expressly not intended to be validly published (Bryant 1996; Cantino et al. 1997; Crane & Kendrick 1997; Kron 1997). These trials suggest that phylogenetic nomenclature has greater potential for stability, universality, and clarity than does traditional nomenclature. Of course, both traditional and phylogenetic nomenclature are subject to instability through changes in circumscription based on new research. Moore (1998) argues, however, that increasing the number of reference taxa in the definition of a name will proportionally multiply the instability caused by such changes compared to that for taxon names referenced by a single type. With regard to accommodating reticulate evolution, Schander (1998b) maintains that phylogenetic nomenclature is more effective than is traditional nomenclature.

These trials, as well as criticisms of the principles, have revealed some problems with the phylogenetic system, as well. If entirely new names are coined for names of clades already recognized in traditional concepts (exemplified in Table 3), there is a loss of historical continuity. Access to the voluminous biological literature of those taxa is cut off, unless some cumbersome

TABLE 3. Example of a phylogenetic taxonomy of the Ericaceae (based on Kron 1997) in which alternate nomenclatural rules are used. In the first case "phylogenetic indicator" suffixes are used, and in the second, traditional names are used.

Using All New Names ("Phylogenetic" Suffixes)	Using Traditional Names (And Suffixes)
Ericalina	Ericales
Enkianthina	Enkianthales
Arbuarctina	Ericanae
Arbutina	Arbutineae
Pyrolina	Pyrolineae
Ericina	Rhodorineae
Bruckenthalina	Cassiopineae
Cassiopina	Cassiopaceae
Callunina	Ericaceae
Phillipina	Ericoideae
Phyllodocina	Phyllodocoideae
Rhodorina	Empetroideae
Empetrina	Empetraceae
Rhododendrina	Rhodoreae
Pentachondrina	Harrimanellineae
Harrimanellina	Harrimanellaceae
Lebetanthina	Lebetanthaceae
Epacridina	Epacridaceae
Gaultheriina	Vaccinioideae
Lyoniina	Lyoniinae
Costerina	Vaccinieae
Pernettyina	Gaultheriinae
Vacciniina	Vacciniinae

cross-reference mechanism is also provided. On the other hand, if names of the traditionally conceived taxa are redefined phylogenetically for the names of the corresponding clades (exemplified in Table 3), the Linnaean ranks will still be connoted. "Old" family names will be nested within other family names, etc., and many users will be confused, especially during a rather lengthy transition period. Names of species are especially problematic. Linnaean binomials include mandatory designation of the genus in the species name, which violates the principles of phylogenetic nomenclature. Due to the sheer numbers of species, uninomials (as normally conceived) are unworkable; however, several authors have suggested means for using specific epithets as uninomials (De Queiroz & Gauthier 1992; Graybeal 1995; Schander 1998b). Another suggestion is to provide each species with a non-Linnaean binomial composed of a rankless forename and the specific epithet. This rankless forename could consist of the the currently accepted genus name, "frozen" into permanent nomenclatural use, or it could consist of any name that is applied to the clade; likewise,the Linnaean binomial could be hyphenated and, thus, considered equivalent to a uninomial (De Queiroz & Gauthier 1992; Graybeal 1995; see comments by Schander 1998b; and Cantino 1998). We fear that confusion between the rankless concept and the Linnaean generic concept will result (see example in Table 4), despite assurances by these authors to the contrary. There are concerns, too, that the attempts to "freeze" concepts of taxa for stability purposes is counterproductive at a time when sufficient data to assess monophyly is just becoming available for a significant number of groups.

Phylogenetic taxonomists that do not view the Linnaean hierarchy to be inherently opposed to the goals of their work attempt to work within the Linnaean system of nomenclature. This school is exemplified by Wiley (1981) who compiled a reasonably encompassing set of conventions in order to retain the valuable aspects of the Linnaean system and historic continuity, while accommodating the demand to recognize only monophyletic taxa (see also Moore 1998).

- The most important of these conventions is phyletic sequencing (Wiley's conventions 2 & 3, p. 205 ff.). The clades occurring in sequence on a pectinate portion of the cladogram are named at the same rank, but the inclusive sister-groups are not named; and the taxa are listed in sequence of their phylogenetic order of origin.
- Fossil taxa are assigned to the category, plesion (convention 7, p. 214 ff), a name of variable rank or, rather, without rank, to indicate sister-group status rather than requiring inflation of the number of ranks.

TABLE 4. Classification of a few selected species and clades of Verbenaceae tribe Lantaneae using phylogenetic taxonomy principles, indention to designate inclusiveness, and the proposed principle of phylogenetic nomenclature to use genus names simply as unranked forenames. The effect of "freezing" the currently accepted combinations to produce the forename-species epithet as binominal designators of species is illustrated.

- Lantaneae
    - Alyosia
    - Lippia
        - Lantana
            - Lippia harleyi
            - Lantana camara
        - Goniostachyum
            - Lippia graveolens
        - Riedelia
            - Lantana brasiliensis
            - Lippia lupulina
        - Zappania
            - Lippia alba
            - Lippia americana
            - Phyla
                - Phyla nodiflora
                - Lippia dulcis

TABLE 5. Wiley's conventions applied to Kron's (1997) cladogram of Ericaceae (with her Ericales changed in rank to Ericaceae, which is more in line with traditional circumscriptions).

- Ericaceae
    - Enkianthoideae
    - Arbutoideae *sedis mutabilis*
    - Pyroloideae *sedis mutabilis*
    - Ericoideae *sedis mutabilis*
        - Cassiopeae
        - Ericeae *sedis mutabilis*
        - Phyllodoceae *sedis mutabilis*
        - Rhodoreae *sedis mutabilis*
            - Empetrinae
            - Rhodorinae
    - Harrimanelloideae
    - Epacridoideae
    - Vaccinioideae
        - Lyonieae
        - Gaultherieae
        - Vaccinieae

- Clades with unresolved or unknown relationships are annotated as *sedis mutabilis* and *incertae sedis*, respectively (conventions 4 & 5, p. 210 ff), and are placed in the classification where their relationships are known.
- Ancestors, if known, are named and accompany the clade(s) to which they give rise (convention 8, p. 222 ff). This convention obviates Brummitt's concern over telescoping taxa.
- Taxa of hybrid and symbiotic origin are classified at the rank at which the lineages giving rise to them diverge (convention 9, p. 225 ff). Remember that a species of hybrid origin is not biphyletic, as some would argue, because the term does not apply. Even if the species is ancestral to a clade, the clade is monophyletic.

These conventions are applied here to Kron's (1997) classification for illustration (Table 5).

Additional modifications to the Linnaean system have been suggested more recently (see also Moore 1998). These include ideas such as: 1) allowing monotypic sister taxa to be unranked and, hence, nonredundant; 2) allowing intercalation of unranked named clades, including those with informal names; 3) reducing the number of required ranks, especially above family; 4) not requiring inclusion of all taxa within nonmandatory subsidiary taxa just because some are so placed; 5) repealing the use of autonyms; and 6) encouraging the use of character-based names.

ISSUE 4. Increasing isolation of floristic and phylogenetic researchers.

The systematic community's single system of nomenclature and desire for universally accepted classifications for the different groups of organisms creates a tension between systematists who view multipurpose classifications and immediate nomenclatural stability as the driving force in systematics and those who advocate phylogenetic classifications as the key to ultimate stability and use (Stevens 1990). With the advent of DNA characters some 15 years ago, this philosophical divergence has evolved into the increasing specialization and isolation of the floristic/field-oriented systematists and the phylogenetic molecular researchers from one another. Indeed, on the one hand, the field- and herbarium-oriented worker often lacks the training to evaluate the results of the DNA-based phylogenetic studies. On the other, the phylogenetic workers, especially if their training has centered on molecular techniques, often lack the skills or inclination to translate their phylogenetic conclusions into usable classifications. Also, many are reluctant to relate the available morphological data to the clades that emerge from the molecular data.

For example, in 1986 Sytsma and Gottlieb provided unequivocal molecular evidence that the genus *Heterogaura* (Onagraceae) is derived from within *Clarkia,* a hypothesis which had been suggested earlier on morphological grounds (see also Gottlieb & Ford 1996). However, these authors did not make the necessary taxonomic and nomenclatural changes warranted by their data. Instead, it fell upon those preparing the " New Jepson Manual" to actually revise the classification (Lewis & Raven 1992).

Recently, *Leiophyllum* and *Loiseleuria* (Ericaceae) were shown to be species of *Kalmia* on the basis of morphology and two distinct sets of macromolecular data (Kron & King 1996). In this case, the authors are well aware of this problem and are in agreement with our concerns. However, they did not make the needed taxonomic changes in deference to a colleague. Until the necessary nomenclatural changes are made, the systematic community will lack the benefit of having a revised classification for these plants.

We believe that the gulf between the two sides will continue to widen as the new generations of graduate students receive less and less training in the full complement of systematic methods, and as systematics positions are filled by molecular biologists with little training beyond molecular laboratory techniques. The median age of researchers who hold positions allowing them to do floristic studies continues to climb. If this situation, which will cause all areas of systematics to suffer, is to be reversed, communication between the two groups must be facilitated. On both sides, workers must become broadly familiar with each other's work. For the benefit of floristic research, as well as education, results from phylogenetic studies must be made accessible to writers of Floras and traditionally trained teachers of systematics, both of whom must seek to incorporate phylogenetic hypotheses into their work.

IMPLEMENTATION ISSUES

The implementation issues, with respect to producing Floras, consist of the responsibilities of the two diverging groups of workers. Floristic researchers must develop means to publish works based on the most scientifically accurate classifications. Phylogenetic/molecular researchers must publish results in a form available for practical use. The onus for having their phylogenetic conclusions accepted by the wider systematic community and the general public falls, at least in part, on the phylogenetic workers themselves.

Phylogenetic/Molecular Researchers

First, if phylogenetic workers lack the basic skills for developing classifications and applying rules of nomenclature, they should correct this deficiency. To remain ignorant of this area of systematics is irresponsible for anyone identifying themselves with the field. Even if they never publish a classification revision themselves (as in the last recommendation below), they should be familiar with the mechanics of preparing such revisions.

Second, when publishing, phylogenetic workers should make the results accessible to floristic workers. At a minimum, they should electronically submit their results to TreeBASE, a distributed database on the Internet (see below). However, they should also write morphological descriptions and identification keys to the taxa receiving revised circumscriptions (based on molecular synapomorphies). The computer program, MacClade (Madison & Madison 1992), can be used to map morphological characters onto DNA-based cladograms (given the existence of a morphological matrix). When molecular researchers are mindful of this goal of accessibility, they will be less likely to treat morphology superficially. Morphological data (including micromorphology, ultrastructure, anatomy, development, and secondary chemistry) can be very powerful in elucidating phylogeny or giving support to DNA-based groups, especially if not restricted to a few often-published diagnostic features.

Finally, if molecular phylogeneticists are uncomfortable in publishing morphological accounts, they should seek to publish in collaboration with skilled monographers. Together, they would be able to make comprehensive data sets available, both to satisfy phylogeneticists' interests and to meet the needs of writers of Floras (and teachers). An excellent example of this type of collaboration is work in progress on the Ericaceae by Kron and her colleagues, who are preparing a monograph, which includes formal circumscriptions, keys, and a revised classification based on morphology, *rbcL*, and *matK* (Judd et al. 1998; Kron et al. 1998).

Floristic Researchers

We admonish the editors of major floristic projects never to appeal to the weight of authority for the classification they use to organize the publication of the Flora. The editorial committee should be chosen to reflect taxonomic as well as geographic specialties. This should bring together specialists familiar enough with the literature to evaluate phylogenetic publications for broad outlines of classifications. Because a phylogenetic classification covering all vascular plants to the species or even family level does not yet exist, the committee will need to establish a revised classification that reflects well-supported relationships. For example, compare the treatment of the genera and families in the Lamiales in the authoritative but out-dated system of Cronquist (1981) to that in recent cladistic studies (Cantino 1992; Cantino et al. 1997; and citations therein). A scientifically accurate Flora will incorporate the latter in preference over the former. This process will become much easier with the publication of an explicit phylogenetic classification covering all flowering plants (see Angiosperm Phylogeny Group 1998).

Of course, major floristic projects also face the problem of having published fascicles or volumes appear over a multi-year period. Therefore, if the volumes and corresponding editorships are organized and delimited to reflect the current state of phylogenetic knowledge, then later adjustments will be more easily made. As new data comes out, the unfinished volumes can be changed without effecting other volumes. For example, the cladistic position of the Lamiales near or within the Scrophulariales (see Cantino 1992) has been established for about ten years. The editors of a Flora should place the Lamiales in the same volume as the Scrophulariales, and the Boraginaceae in the same volume as the Solanales. As the time approaches for publication of the particular volume, the contributing authors would be able to provide the latest phylogenetic classification of the included taxa without difficulty, and references to these families in the introductory volumes should be minimally outmoded. There would still be those unusual cases, such as the Plumbaginaceae, which have had disparate, equivocal placements based on morphology. If this family had been planned for inclusion near the Primulales but later verified as cladistically close to the Polygonales (Lledó et al. 1998), the new findings should be summarized in the appropriate volume that appears after the discovery. Thus, careful editorial planning gains much in long term usefulness.

These problems are really a product of publication technologies that are rapidly becoming obsolete. As floristic projects become increasingly oriented toward electronic publication and frequent updating of materials, the problems of dealing with new findings are greatly minimized. The future prospects of frequent updating also make a case for improved nomenclatural rules to prevent a clade's name changing with rank.

AN URGENT NEED

Once the editors of a floristic project have committed to an up-to-date phylogenetic classification and organization, the greatest problem is making sure that the contributing authors have access to published data and can effectively use those data. The answer to this need is a database of published cladistic analyses that covers vascular plants, bryophytes, and green algal groups on a world-wide basis. (Databases for fungi, animals, and various protist clades are also needed.)

Donoghue and his associates have previously recognized the urgency of this need. (Sanderson et al., 1993) inventoried 76 journals for three years and found 882 phylogenetic analyses and an annual increase of publication rate of over 15%. About 25% dealt with angiosperms. As a consequence of the divergence of phylogenetic studies from the realm of the morphologically trained systematist, the majority of the analyses were published in new journals covering molecular biology. In responding to this important need, this research group, with the institutional backing of Harvard University, the University of California at Davis, and the National Science Foundation, designed the relational database, TreeBASE, for access on the Internet (Donoghue et al, 1994). With only a total of 320 included studies (mostly angiosperms, fungi, and arachnids), TreeBASE is still in the pilot stage. It promises to provide the resources for which we are calling. Currently one may view and manipulate published cladograms and data matrices. Because the process of entering data from published papers is labor intensive, authors with papers in press are encouraged to submit data matrices and trees in digital form directly to TreeBASE.

The Tree of Life web site (Maddison and Maddison 1996), a distributed database on the Internet, is a secondary source, especially for botanists unfamiliar with cladistic techniques. However it is not a comprehensive index of the phylogenetic literature and is designed more as a educational resource for the general public.

We hope that TreeBASE will continue to be supported and come to include all available phylogenetic studies of vascular plants, as we envision for the recommended database. However, the information available through such a phylogenetic database would be unfamiliar to a significant number of floristic contributors who are less familiar with the molecular and/or cladistic literature and techniques. Therefore, editors would also be responsible for assisting these contibutors to understand the methods and to use effectively the cladistic literature.

CONCLUSIONS

To incorporate the available, well-supported phylogenetic data into Floras, especially large floristic projects, which will serve as the basis of national and international standards of taxonomic concepts and usage, we offer three major recommendations based on the issues raised. The first and most urgent is the continued support and expansion of TreeBASE, the Internet-accessible relational database on published cladistic (morphological and molecular) analyses. Based on the information accessible through this database, the editors and writers of Floras should make scientific accuracy the primary consideration. This will require careful planning and knowledgeable, scientifically bold leadership in evaluating and adopting phylogenetic classifications as new data become available. It will also require an investment in training the contributors in interpreting cladistic and molecular research. Finally, there must be increased communication and collaboration between molecular and field-oriented systematists, to make phylogenetic results accessible for practical use. To facilitate this communication, cladistic researchers are encouraged to submitted their data and results to TreeBASE.

ACKNOWLEDGMENTS

We thank the Botanical Research Institute of Texas for the use of library and office facilities during the preparation of this paper. Discussions with numerous colleagues contributed to the thoughts expressed here. Peter F. Stevens, Keven De Queiroz, and Richard Brummitt kindly supplied us with reprints of their publications. Dr. Kathleen Kron reviewed the completed manuscript.

REFERENCES

ADAMS, C.D. 1972. Flowering plants of Jamaica. University of the West Indies, Mona, Jamaica.

ANGIOSPERM PHYLOGENY GROUP. 1998. An ordinal classification for the families of flowering plants. Ann. Missouri Bot. Gard. 85:531–553.

BACKLUND, A. and K. BREMER. To be or not to be—principles of classification and monotypic plant families. Taxon 47:391–400.

BRUMMITT, R.K. 1996. In defense of paraphyletic taxa. In: Maesen, L.J.G. van der, Burgt, X. M. vand der, and Medenbach de Rooy, J.M. van, eds. The biodiversity of African plants. Proceddings XIVth AETFAT Congress 22–27 August 1994. Dordrecht, Wageningen, the Netherlands.

________. 1997. Taxonomy versus cladonomy, a fundamental controversy in biological systematics. Taxon 46:723–734. Pp. 371–384.

________ and M.S.M. SOSEF. 1998. Paraphyletic taxa are inherent in Linnaean classification—a reply to Freudenstein. Taxon 47:411–412.

BRYANT, H.N. 1994. Comments on the phylogenetic definition of taxon names and conventions regarding the naming of crown clades. Syst. Biol. 43:124–130.

________. 1996. Explicitness, stability, and universality in the phylogenetic definition and usage of taxon names: A case study of the phylogenetic taxonomy of the Carnivora (Mammalia). Syst. Biol. 45:174–189.

CANTINO, P.D. 1992. Evidence for a polyphyletic origin of the Labiatae. Ann. Missouri Bot. Gard. 79:361–379.

________. 1998. Binomials, hyphenated uninomials and phylogenetic nomenclature. Taxon 47:425–430.

________, R.G. OLMSTEAD and S.J. WAGSTAFF. 1997. A comparison of phylogenetic nomenclature with the current system: A botanical case study. Syst. Biol. 46:313–331.

CORRELL, D.S. and M.C. JOHNSTON. 1970. Manual of the vascular plants of Texas. Texas Research Foundation, Renner, Texas.

CRANE, P.R. and P. KENRICK. 1997. Problems in cladistic classification: Higher-level relationships in land plants. Aliso 15:87–104.

CRONQUIST, A. 1981. An integrated system of classification of flowering plants. Columbia University Press, New York.

________. 1987. A botanical critique of cladism. Bot. Rev. 53:1–52.

DAVIS, J.I. and K.C. NIXON. 1992. Populations, genetic variation and the delimitation of phylogenetic species. Syst. Biol. 41:421–435.

DE QUEIROZ, K. 1996. A phylogenetic approach to biological nomenclature as an alternative to the Linnaean systems in current use. In: J.L. Reveal, ed. Proceedings of a mini-symposium on biological nomenclature in the 21st century. University of Maryland. Published as on-line text, Internet address: www.life.umd.edu/bees/96sym.htm

________. 1997. The Linnaean hierarchy and the evolutionization of taxonomy, with emphasis on the problem of nomenclature. Aliso 15:124–144.

________ and J. GAUTHIER. 1990. Phylogeny as a central principle in taxonomy: Phylogenetic definitions of taxon names. Syst. Zool. 39:307–322.

________ and ________. 1992. Phylogenetic taxonomy. Annual Rev. Ecol. Syst. 23:449–480.

________ and ________. 1994. Toward a phylogenetic system of biological nomenclature. Trends Ecol. Evol. 9:27–31.

DONOGHUE, M.J. and P.D. CANTINO. 1988. Paraphyly, ancestors, and the goals of taxonomy: A botanical defense of cladism. Bot. Rev. 54:107–128.

________, M.J. SANDERSON, W. PIEL, and T. ERIKSSON. 1994. TreeBASE: A database of phylogenetic knowledge. Internet address: http://www.herbaria.harvard.edu/treebase/

FLORA OF NORTH AMERICA EDITORIAL COMMITTEE. 1993+. Flora of North America north of Mexico. (to be published in multiple volumes) Oxford Univ. Press, New York.

FREUDENSTEIN, J.V. 1998. Paraphyly, ancestors, and classification—response to Sosef and Brummitt. Taxon 47:95–104.

GOTTLIEB, L.D. and V.S. FORD. 1996. Phylogenetic relationships among the sections of *Clarkia* (Onagraceae) inferred from the nucleotide sequences of *PgiC*. Syst. Bot. 21:45–62.

GRAYBEAL, A. 1995. Naming species. Syst. Biol. 44:237–250.

HUMPHRIES, C.J. and J.A. CHAPPILL. 1988. Systematics as science: A response to Cronquist. Bot. Rev. 54:129–144.

JUDD, W.S., C.S. CAMPBELL, E.A. KELLOGG, and P.E. STEVENS. 1999. Plant systematics: A phylogenetic approach. Sinauer Associates, Sunderland, MA.

JUDD, W.S., K.A. KRON, A.A. ANDERBERG, D.M. CRAYN, C.J. QUINN, and P.F. STEVENS. 1998. Major clades of Ericaceae. I: Morphology. Amer. J. Bot. 65 (6, suppl.):138–139.

________, R.W. SANDERS, and M.J. DONOGHUE. 1994. Angiosperm family pairs: Preliminary phylogenetic analyses. Harv. Pap. Bot. 1(5)–51.

KRON, K.A. 1997. Exploring alternative systems of classification. Aliso 15:105–112.

________, W.S. JUDD, A. A. ANDERBERG, D.M. CRAYN, R. FULLER, P.A. GADEK, J.L. LUTEYN, C. J. QUINN and P.F. STEVENS. 1998. Major clades of Ericaceae. II: Molecular evidence. Amer. J. Bot. 65 (6, suppl.):139–140.

________ and J.M. KING. 1996. Cladistic relationships of *Kalmia, Leiophyllum*, and *Loiseleuria* (Phyllodoceae, Ericaceae) based on *rbcL* and nrITS data. Syst. Bot. 21:17–29.

LEWIS H. and P.H. RAVEN. 1992. New combinations in the genus *Clarkia* (Onograceae). Madroño 39:163–169.

LIDÉN M., B. OXELMAN, A. BACKLUND, L. ANDERSSON, B. BREMER, R. ERIKSSON, R. MOBERG, I. NORDAL, K. PERSSON, M. THULIN, and B. ZIMMER. 1997. Charlie is our darling. Taxon 46:735–738.

MADDISON, D.R. AND W.P. MADDISON. 1992. MacClade: Interactive analysis of phylogeny and character evolution, version 3.04. Sinauer Associates, Sunderland, Mass.

_______ and _______. 1996. The tree of life: A distributed Internet project containing information about phylogeny and biodiversity. Internet address: http://phylogeny.arizona.edu/tree/phylogeny.html

MOORE, G. 1998. A comparison of traditional and phylogenetic nomenclature. Taxon 47:561-580.

MORELL, V. 1996. TreeBASE: the roots of phylogeny. Science 273:569.

SANDERSON, M.J., B.G. BALDWIN, G. BHARATHAN, C.S. CAMPBELL, C. VON DOHLEN, D. FERGUSON, J.M. PORTER, M.F. WOJCIECHOWSKI, and M.J. DONOGHUE. 1993. The growth of phylogenetic information and the need for a phylogenetic data base. Syst. Biol. 42:562–568.

_______, M.J. DONOGHUE, W. PIEL, and T. ERIKSSON. 1994. TreeBASE: a prototype database of phylogenetic analyses and an interactive tool for browsing the phylogeny of life. *Amer. J. Bot.* 81:183

SCHANDER, C. 1998a. Types, emendations and names–a reply to Lidén & al. Taxon 47:401–406.

_______. 1998b. Mandatory categories and impossible hierarchies—a reply to Sosef. Taxon 47:407–410.

SOSEF, M.S.M. 1997. Hierarchical models, reticulate evolution and the inevitability of paraphyletic supraspecific taxa. Taxon 46:75–85.

STEVENS, P.F. 1990. Nomenclatural stability, taxonomic instinct, and flora writing—a recipe for disaster? In: P. Baas, K. Kalkman, and R. Geesink, eds. The Plant Diversity of Malesia: Proceedings of the Flora Malesiana Symposium, 1989. Kluwer Academic Publishers, Dordrecht, Netherlands. Pp. 387–410.

_______. 1997. How to interpret botanical classifications—suggestions from history. BioScience: 47:243–250.

_______. 1998. What kind of classification should the practising taxonomist use to be saved? In: J. Dransfield, M.J.E. Coode, and D.A. Simpson, eds. Plant Diversity in Malesia III: Proceedings of the third international Flora Malesiana Symposium, 1995. Royal Botanic Gardens, Kew. Pp. 295–319.

STUESSY, T. F. 1990. Plant taxonomy: The systematic evaluation of comparative data. Columbia Univ. Press, New York.

_______. 1997. Classification: More than just branching patterns of evolution. Aliso 15:113–124.

SYTSMA, K.J. and L.D. GOTTLIEB. 1986. Chloroplast DNA evidence for the origin of the genus *Heterogaura* from a species of *Clarkia* (Onograceae). Proc. Natl. Acad. Sci. USA 83:5554–5557.

UHL, N. and J. DRANSFIELD. 1987. Genera palmarum. Allen Press, Lawrence, Kansas.

WAGNER, W.H., JR. and J.M. BEITEL. 1992. Generic classification of modern North American Lycopodiaceae. Ann. Missouri Bot. Gard. 79:676–686.

WELZEN, P.C. VAN. 1997. Paraphyletic groups or what should a classification entail. Taxon 46:99–103.

_______. 1998. Phylogenetic versus Linnaean taxonomy, the continuing story. Taxon 47:413–424.

WILEY, E.O. 1981. Phylogenetics: The theory and practice of phylogenetic systematics. John Wiley & Sons, New York.

WOOD, C.E., JR. 1958. The genera of the woody Ranales in the southeastern United States. J. Arnold Arbor. 39:296–346.

DATABASE RESOURCES IN QUATERNARY PALEOBOTANY

STEPHEN T. JACKSON[1]
ERIC C. GRIMM[2]
ROBERT S. THOMPSON[3]

[1]*Department of Botany*
Aven Nelson Building
University of Wyoming
Laramie, WY 82071, U.S.A.

[2]*Illinois State Museum*
Research and Collections Center
1920 South 10½ Street
Springfield, IL 62703, U.S.A.

[3]*United States Geological Survey*
Box 25064, Mail Stop 919
Denver Federal Center
Denver CO 80225, U.S.A.

ABSTRACT

Studies of pollen and plant macrofossils from Quaternary sediments provide information on past flora, vegetation, climate, and phytogeography. Quaternary paleobotanical data are now being compiled and curated in cooperative databases, supported by funding from the Paleoclimatology Program of the National Oceanic and Atmospheric Administration (NOAA). These data are being permanently archived in electronic format at the World Data Center-A for Paleoclimatology, located at the NOAA National Geophysical Data Center (NGDC) in Boulder, Colorado. Three North American databases are being developed. The North American Pollen Database (NAPD) archives Quaternary pollen assemblages from throughout the United States and Canada. This database is relatively mature and can be readily accessed through the NGDC web server (*http://www.ngdc.noaa.gov/paleo/paleo.html*). The database has visualization software, including SiteSeer, a program that allows selection and extraction of pollen diagrams from sites in the database, and ShowTime, a program that displays time-series maps of pollen percentages for eastern and central north America. The North American Plant Macrofossil Database (NAPMD) archives plant macrofossil data from "wet" sediments in the United States, Canada, and Greenland. The NAPMD is approaching completion. A complete site inventory is available now on the NGDC web server. Efforts are underway to complete the database and to display the data using SiteSeer and ShowTime. The Western North American Midden Database (WNAMD) compiles plant-macrofossil data from fossil packrat middens in arid portions of western United States and Canada. This database is under development, but a site and taxon inventory is available electronically.

RESUMEN

Los estudios de polen y macrofósiles de plantas en sedimentos del Cuaternario dan información de la flora, vegetación, clima y fitogeografía del pasado. Los datos paleobotánicos del Cuaternario se están compilando y manteniendo en bases de datos cooperativas, mantenidas con fondos del Programa de Paleoclimatología de la National Oceanic and Atmospheric Administration (NOAA). Estos datos se archivan permanentemente en formato electrónico en el World Data Center-A para la Paleoclimatología, situado en el NOAA National Geophysical Data Center (NGDC) en Boulder, Colorado. Se están desarrollando tres bases de datos norteamericanas. La base de datos del polen de Norteamérica (NAPD) archiva grupos de polen del cuaternario de los Estados Unidos y Canadá. Esta base de datos esta relativamente "madura" y puede accederse a ella mediante el servidor de la NGDC (*http://www.ngdc.noaa.gov/paleo/paleo.html*). La base de datos tiene software de visualización, incluyendo SiteSeer, un programa que permite la selección y extracción de diagramas polínicos de la base de datos, y ShowTime, un programa que visualiza mapas de series temporales de porcentajes de polen del este y centro de Norteamérica. La base de datos de macrofósiles de plantas de Norteamérica (NAPMD) archiva datos de macrofósiles de planta procedentes de sedimentos "húmedos" en Estados Unidos, Canadá y Groenlandia. La NAPMD está cerca de ser completada. En la actualidad está disponible un inventario completo del sitio en el servidor web de la NGDC. Se están realizando esfuerzos para completar la base de datos y visualizar los datos usando SiteSeer y ShowTime. La base de datos Western North American Midden (WNAMD) compila datos de macrofósiles vegetales de yacimientos de ratas de bosque en partes áridas del oeste de Estados Unidos y Canadá. Esta base de datos está en desarrollo, pero está disponible electrónicamente un sitio y un inventario de taxa.

SIDA, Bot. Misc. 18: 113–120. 2000

INTRODUCTION

During the past forty years, paleoecological studies of pollen and/or plant macrofossils from Quaternary sediments have been done in nearly every part of North America. Because these data are of considerable value in inferring past climate changes and in testing the climate models that are being used to predict future climate changes, a coordinated international effort is underway to archive these and other paleoenvironmental data in publicly accessible, digital format at the World Data Center-A (WDC-A) for Paleoclimatology (Webb et al. 1994; Anderson 1995). Efforts in North America are sponsored by the National Oceanic and Atmospheric Administration (NOAA) Paleoclimatology Program and the National Science Foundation (NSF) Earth System History Program. The WDC-A for Paleoclimatology is located at the NOAA National Geophysical Data Center (NGDC) in Boulder, Colorado.

Applications of Quaternary paleoecological data are not limited to paleoclimatology. Paleoecological studies are providing a wealth of information on the vegetational, floristic, and biogeographic history of North America (Betancourt et al. 1990; Webb et al. 1993, Jackson et al. 1997). Paleoecological data can be used to test biogeographic hypotheses, document the history of plant populations, and determine the antiquity of plant communities (e.g., Thompson 1988; Webb 1988; Betancourt et al.1991; Miller 1992, 1993; Nowak et al. 1994; Baker et al. 1996; Jackson & Singer 1997). Some recent studies of particular relevance to the floristic history of the southeastern United States include Watts et al. (1992), Grimm et al. (1993), Kneller and Peteet (1993, 1999), Jackson and Givens (1994), Jackson and Weng (1999), and Jackson et al. (2000).

The databases archived at the WDC-A for Paleoclimatology can be accessed by anyone (*http://www.ngdc.noaa.gov/paleo/paleo.html*), and represent an important resource for floristic botanists, systematists, biogeographers, and ecologists. In this paper we describe the three North American paleobotanical databases, and provide information on their goals, structure, current status, and applications. The databases include the North American Pollen Database (NAPD), the North American Plant Macrofossil Database (NAPMD), and the Western North American Packrat Midden Database (WNAPMD). Each database includes unique kinds of data, but all share fundamental database structures and are designed to be intercompatible.

NORTH AMERICAN POLLEN DATABASE

Pollen accumulated in sediments of lakes, bogs, caves, and other depositional environments provide information about past flora and vegetation at local to regional scales. Spatial resolution can be limited by the dispersibility of pollen grains in the atmosphere, and taxonomic resolution is often limited to genus or sometimes family (e.g., Poaceae, Cyperaceae). Nevertheless, pollen data provide reliable information on past vegetation, and interpretation of the fossil record is facilitated by the existence of thousands of modern pollen assemblages from surface sediments and pollen traps. Issues of resolution and interpretation are discussed in recent reviews (Webb 1993; Sugita 1994; Jackson 1994, 1997). Pollen data are usually expressed as percentages of a sum which typically includes all terrestrial pollen taxa in the assemblages.

The North American Pollen Database (NAPD) archives pollen data from both modern and fossil assemblages. The database effort is being directed by E.C. Grimm at the Illinois State Museum. An advisory committee has guided development of the database; this committee includes K.J. Gajewski (Univ. Ottawa), G.L. Jacobson, Jr. (Univ. Maine), G.M. MacDonald (Univ. California - Los Angeles), L.J. Maher (Univ. Wisconsin), V. Markgraf (Univ. Colorado), P.J.H. Richard (Univ. Montrèal), T. Webb III (Brown Univ.), and C. Whitlock (Univ. Oregon). John Keltner was database technician during the early development of NAPD (1990–1995), and since 1995 has been coordinating paleoclimate database activities for NGDC, including all three databases discussed in this paper. Stephen Porter has been NAPD database technician since 1995.

At the time of this writing, fossil pollen data from 907 entities (cores, exposures, etc.) and modern pollen data from 1829 surface samples were archived in the NAPD (Fig. 1). This number is growing steadily as more sites are added to the database. The fundamental data consist of

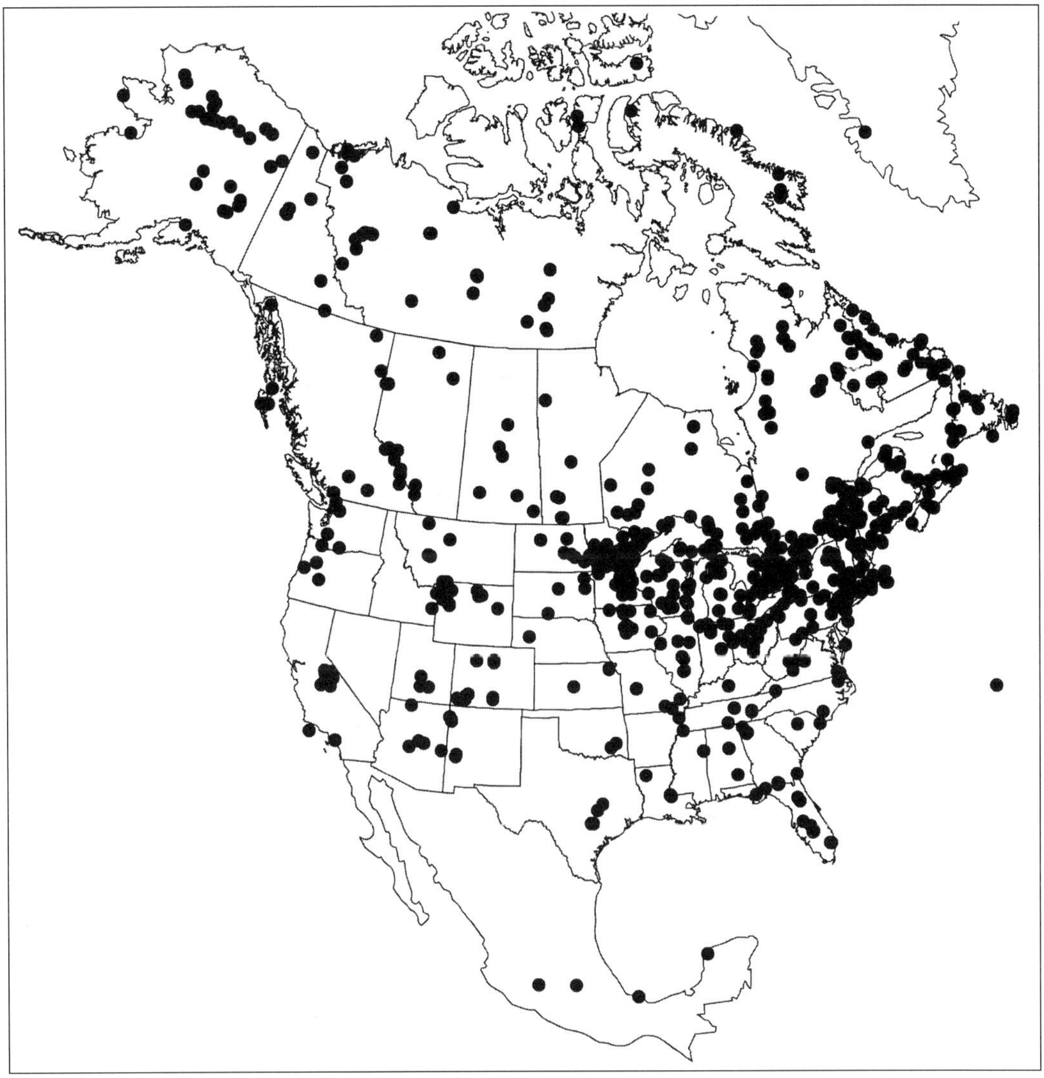

FIG. 1. Map of sites currently in the North American Pollen Database.

raw pollen counts for each sample from each site. The database also includes metadata on each site (e.g., geographic coordinates, elevation, site type, investigator, literature citation), and on site chronology (e.g., radiocarbon and other dates, age-model, age-depth table).

Data can be obtained from NAPD in one of several forms. First, the base tables (consisting of all raw data for all taxa) can be obtained in Paradox 4.5 format. Second, raw pollen counts for the 70 most important pollen taxa in the Northern Hemisphere can be obtained for individual entities in either ASCII files or TILIA spreadsheets. The latter are spreadsheets designed specifically for pollen data management; they can be readily uploaded into EXCEL, Quattro-Pro, or other spreadsheet formats. Information concerning TILIA can be obtained at *http://www.ngdc.noaa.gov/paleo/tilia.html*). Finally, pollen percentages for the fifteen most common pollen types at each site can be extracted as ASCII files.

The NAPD has several additional features of interest. First, it has a search engine that permits the user to locate files corresponding to particular sites, geographic coordinates, political entities (e.g., states/provinces), investigators, etc. This engine can be very useful in finding and retrieving data from specific areas and/or time periods of interest. Second, the NAPD has a companion program, SiteSeer, which allows retrieval and display of pollen diagrams from specific sites. SiteSeer displays a map of North America showing locations of sites in the NAPD. By clicking on an individual site, the user can retrieve metadata for the site, and can display a pollen diagram of the 15 most common pollen types at that site. The program also includes a search feature. Diagrams and tables of pollen percentages can be printed directly from the screen display, or captured for export to other programs. ShowTime, another NAPD-based program, displays time-series maps (15,000 to 0 yr B.P.) of pollen abundances for 16 pollen types in North America.

The NAPD home page is *http://www.ngdc.noaa.gov/paleo/napd.html.* SiteSeer and ShowTime can be downloaded from *http://www.ngdc.noaa.gov/paleo/softlib.html.* The NAPD is now part of a global effort at developing pollen data. The Global Pollen Databases (GPD) can be accessed at *http://www.ngdc.noaa.gov/paleo/gpd.html.* The GPD also include data from Europe, Latin America, and parts of Asia; data from other parts of the world (Africa, Australia, India, IndoMalesia) will eventually be incorporated as well.

NORTH AMERICAN PLANT MACROFOSSIL DATABASE

Plant macrofossils (seeds, fruits, foliage, wood, bud scales, megaspores, etc.) preserved in sediments of lakes, wetlands, and other basins provide information about past flora and vegetation at local scales, but can be applied toward reconstructing floristic and vegetational patterns at regional and subcontinental scales (Jackson et al. 1997). Plant macrofossils are poorly dispersed in the atmosphere and hence provide records of local vegetation (excepting obvious cases of fluvial sediments in large catchments). Morphology and anatomy of plant organs frequently allows identification to the species level, so plant macrofossils often offer greater taxonomic resolution than pollen data. Issues of interpretation of macrofossil records from sediments of lakes and other wet environments are reviewed by Dunwiddie (1987), Miller (1990), Jackson (1997), and Jackson et al. (1997). Data are presented in a variety of ways, including presence/absence, counts per unit volume of sediment, and percentages of a specified macrofossil sum.

The North American Plant Macrofossil Database (NAPMD) effort is being directed by S.T. Jackson at University of Wyoming. Development of the NAPMD has been guided by an advisory committee consisting of R.S. Anderson (Northern Arizona Univ.), R.G. Baker (Univ. Iowa), P.J. Bartlein (Univ. Oregon), E.C. Grimm (Illinois State Museum), A. LaRouche (Univ. Montrèal), J.H. McAndrews (Royal Ontario Museum), N.G. Miller (New York Biological Survey), and R.S. Thompson (U.S. Geological Survey). Database technicians involved in the effort have included D.K. Singer (1995–1997), T.W. Chumley (1997–1998), and Sharon Stewart (since 1998).

The NAPMD is designed to archive and make available plant macrofossil data from Quaternary sediments of the United States, Canada, and Greenland. Highest priority is placed on data from Late Quaternary sediments (0-30,000 yr B.P.), but older assemblages are being incorporated as well, including some from the late Pliocene. The data include presence/absence and counts (e.g., number per unit-volume of sediment) of macroscopic plant organs. Pollen, spores, stomata, and dispersed cuticle are not included, with the exceptions of megaspores (e.g., *Selaginella*) and oöspores (Charophyta). Taxonomic domain includes all members of Kingdom Plantae (i.e., all tracheophytes, bryophytes, hepatophytes, and anthocerophytes), plus Charophyta. The NAPMD is currently restricted to sites and deposits dominated by physical depositional processes. Examples include lakes, wetlands, buried soils, estuaries, fluvial, glacial, and aeolian deposits, and caves. Exposed, radiocarbon-dated logs and stumps, which are especially common at high latitudes and high elevations, are being incorporated. Macrofossils from archeological contexts are not being included. Macrofossils from packrat middens are also excluded; those are being compiled in a separate database (see below).

The database is being organized following structures and definitions developed by the NAPD. So far, we have identified 1073 entities from 622 sites. To date, we have compiled data from 649 entities representing 408 sites (Fig. 2). Of these, 106 have been contributed directly by the analysts, and 543 have been extracted from the literature (transcription of tables or digitizing of diagrams).

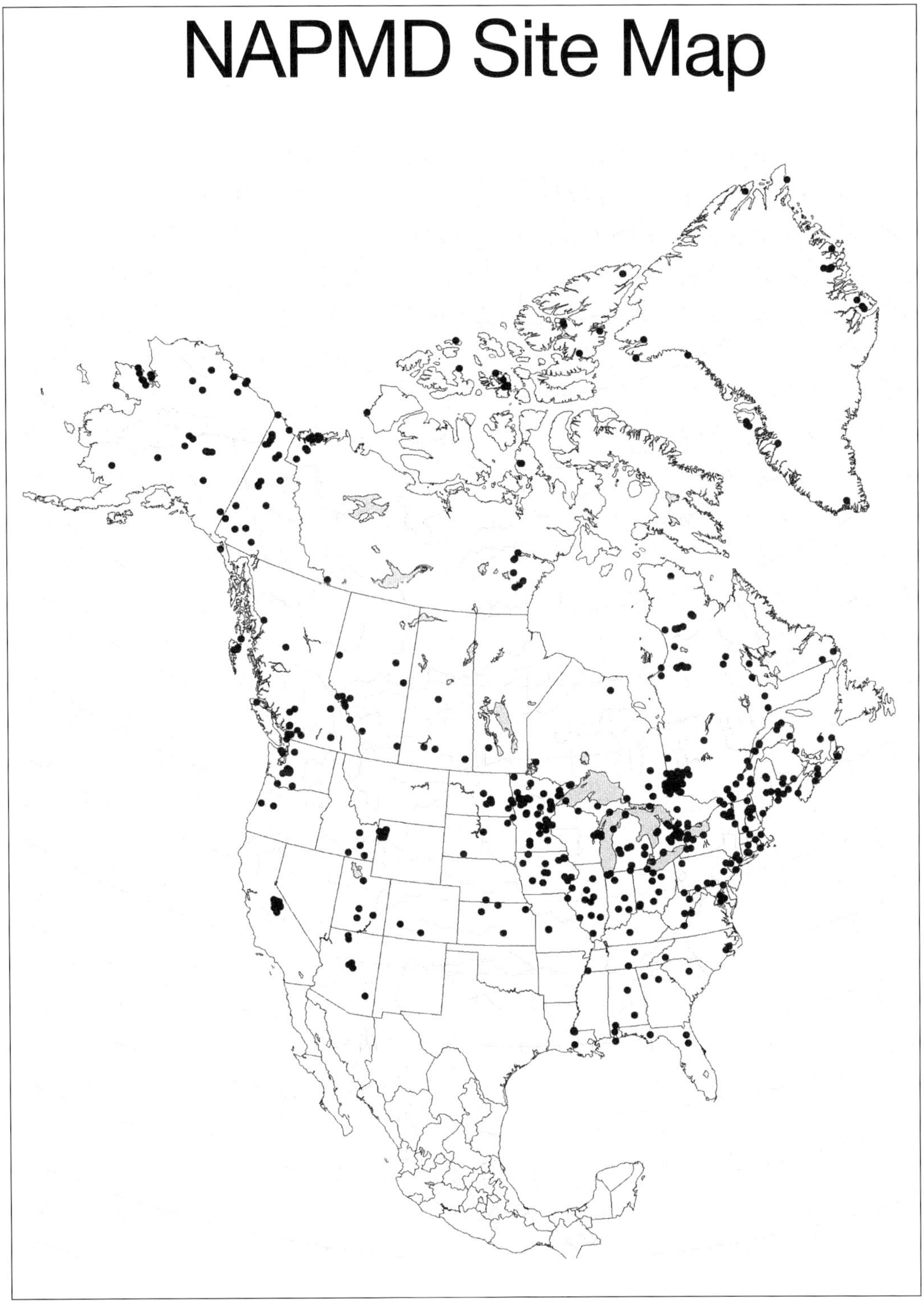

FIG. 2. Map of sites currently in the North American Plant Macrofossil Database Inventory. Data from most of these sites have been acquired and are being incorporated into the database.

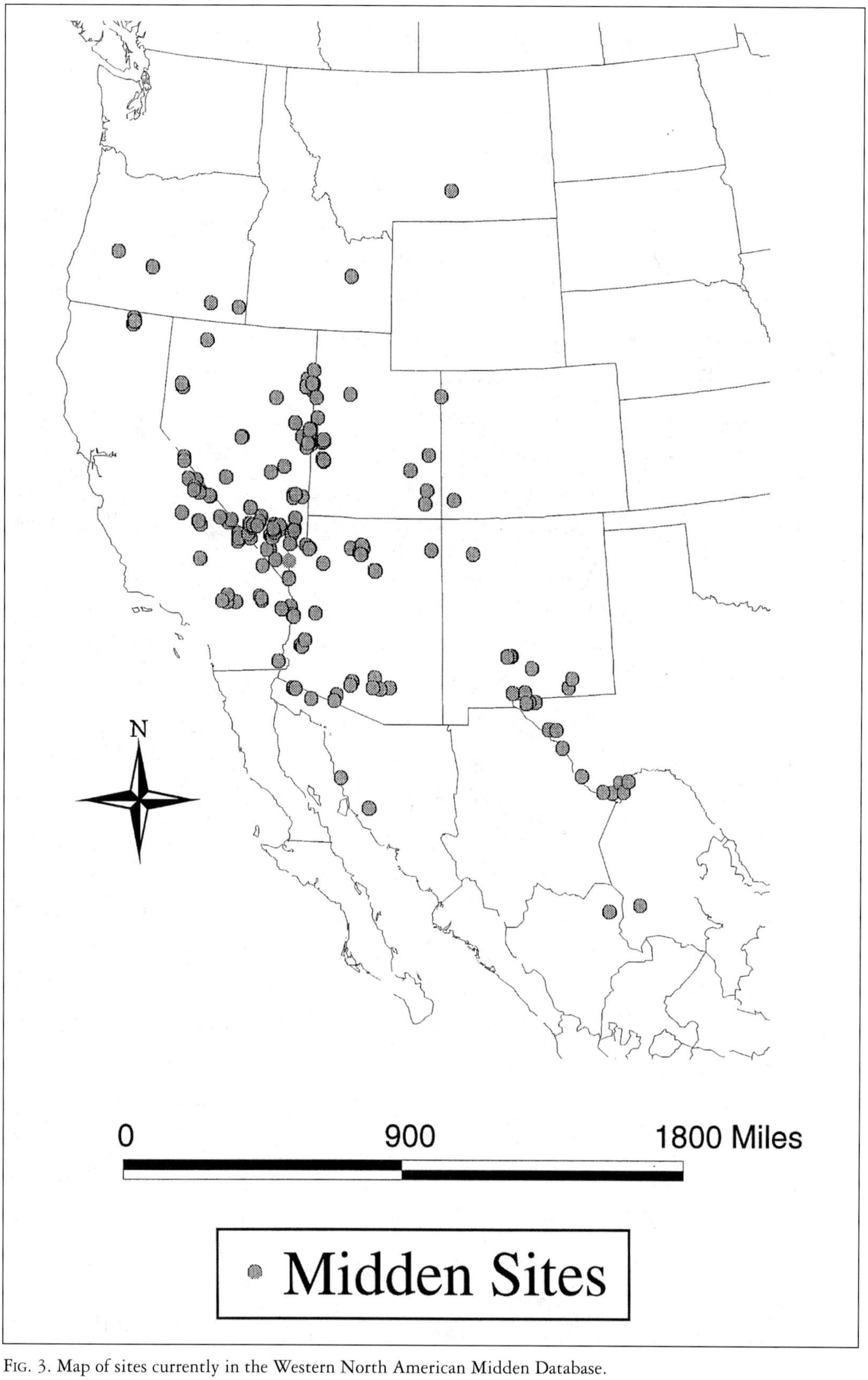

FIG. 3. Map of sites currently in the Western North American Midden Database.

The NAPMD is currently not as advanced as the NAPD in terms of availability and associated programs. A site inventory, which includes metadata (name, geographic coordinates, elevation, site type, age range, investigator, literature citation) for all macrofossil sites in North America and Greenland that we have identified to date, is available at the NAPMD web site (*http://www.uwyo.edu/bot/NAPMD/napmd.htm*), which is linked to the NGDC paleo web page. The full database should be available on the web in late 1999; interested readers should check the NAPMD home page or the NGDC paleo web page for information and updates. Macrofossil data in the NAPMD will be available as Paradox tables, TILIA files, or ascii files. Efforts at developing a search engine (similar to the one used in NAPD), a MacroSiteSeer program, and a MacroShowTime (to display time-series of presence/absence maps (e.g., Jackson *et al.* 1997) for selected taxa) are underway.

WESTERN NORTH AMERICAN MIDDEN MACROFOSSIL DATABASE

Analysis of plant macrofossils from fossil packrat middens during the past two decades has revolutionized our understanding of the Late Quaternary floristic, vegetational, and climatic history of arid regions of western North America (Thompson 1988; Betancourt et al. 1990). Packrats of the genus *Neotoma* collect plant materials in the vicinity of their nests, which are usually in caves or crevices of bedrock escarpments. The plant materials are incorporated into middens in the nests, where they can be preserved for thousands of years provided the material remains dry. Plant materials from the middens can be radiocarbon-dated and identified. The midden macrofossil assemblages provide a record of local flora and vegetation, with taxonomic precision typically at the species level. Midden assemblages have been recovered and studied from as far north as south-central British Columbia, and throughout the western United States (Betancourt et al. 1990). Spaulding et al. (1990) provide a detailed discussion of methodology and interpretation of midden assemblages. The abundance of well-preserved plant tissues in middens has led to isotopic and anatomical studies aimed at understanding evolutionary and physiological responses to climatic change (Van de Water et al. 1994; Hunter et al. 1999).

The Western North American Packrat Midden Database (WNAMD) effort is directed by R.S. Thompson at the United States Geological Survey. Katherine H. Anderson has served as database technician for this effort. To date, the WNAMD has compiled macrofossil records from 939 packrat middens in the United States and Mexico (Fig. 3). Like the NAPMD, the WNAMD is still under development. Interested readers should check the WNAMD web page (*http://climchange.cr.usgs.gov/data/midden/*) for status and updates. This web site currently has a search engine which allows the user to query the database by site, author, region, time period, or taxon. The search engine would be particularly valuable to users interested in learning what is known about past occurrences of a particular species or the floristic history of a specific region.

ACKNOWLEDGMENTS

Development of the databases described in this paper has been supported by grants from the National Oceanic and Atmospheric Administration (NOAA) Paleoclimatology Program. Sharon Stewart, Tim Chumley, and Kathy Anderson assisted in preparing the figures. We thank the many individuals who have contributed to the database efforts.

REFERENCES

ANDERSON, D.M. 1995. Global paleoenvironmental data. PAGES Workshop Report Series 95–2.

BAKER, R.G., E.A. BETTIS III, D.P. SCHWERT, D.G. HORTON, C.A. CHUMBLEY, L.A. GONZALES, and M.K. REAGAN. 1996. Holocene paleoenvironments of northeast Iowa. Ecol. Monogr. 66:203–234.

BETANCOURT, J.L., T.R. VAN DEVENDER, and P.S. MARTIN. 1990. Packrat middens: the last 40,000 years of biotic change. University of Arizona Press, Tucson.

BETANCOURT, J.L., W.S. SCHUSTER, J.B. MITTON, and R.S. ANDERSON. 1991. Fossil and genetic history of a pinyon pine (*Pinus edulis*) isolate. Ecology 72:1685–1697.

DUNWIDDIE, P.W. 1987. Macrofossil and pollen representation of coniferous trees in modern sediments from Washington. Ecology 67:58–68.

GRIMM, E.C., G.L. JACOBSON, Jr., W.A. WATTS, B.C.S. HANSEN, and K.A. MAASCH. 1993. A 50,000-year record of climate oscillations from Florida and its temporal correlation with the Heinrich events. Science 261:198–200.

HUNTER, K.L., B.R. RIDDLE, J.L. BETANCOURT, K.L. COLE, T.R. VAN DEVENDER, and W.G. SPAULDING. 1999. Polyploidy changes during the past 18,000 years in the North American desert shrub *Larrea tridentata*. Nature (in review).

JACKSON, S.T. 1994. Pollen and spores in Quaternary lake sediments as sensors of vegetation composition: theoretical models and empirical evidence. In: A. Traverse, ed. Sedimentation of organic particles. Cambridge University Press. Pp. 253–286.

JACKSON, S.T. 1997. Documenting natural and human-caused plant invasions using paleoecological methods. In: J.O. Luken and J.W. Thieret, eds. Assessment and management of plant invasions. Springer. Pp. 37–55

JACKSON, S.T. and C.R. Givens. 1994. Late Wisconsinan vegetation and environment of the Tunica Hills region, Louisiana/Mississippi. Quat. Res. 41:316–325.

JACKSON, S.T., J.T. Overpeck, T. WEBB III, S.E. KEATTCH, and K.H. ANDERSON. 1997. Mapped plant-macrofossil and pollen records of late Quaternary vegetation change in eastern North America. Quat. Sci. Rev. 16:1–70.

JACKSON, S.T. and D.K. Singer. 1997. Climate change and the development of Coastal Plain disjunctions in the central Great Lakes region. Rhodora 99:101–117.

JACKSON, S.T., R.S. WEBB, K.H. ANDERSON, J.T. OVERPECK, T. WEbb III, J.W. WILLIAMS, and B.C.S. HANSEN. 2000. Vegetation and environment in eastern North America during the last glacial maximum. Quat. Sci. Rev. 19:409–508.

JACKSON, S.T., and C. Weng. 1999. Late Quaternary extinction of a tree species in eastern North America. Proc. Natl. Acad. Sci. 96:13847–13852.

KNELLER, M. and D.M. PETEET. 1993. Late-Quaternary climate in the Ridge and Valley of Virginia, U.S.: changes in vegetation and depositional environment. Quat. Sci. Rev. 12:613–628.

KNELLER, M. and D. PETEET. 1999. Late-glacial to early Holocene climate changes from a central Appalachian pollen and macrofossil record. Quat. Res. 51:133–147.

MILLER, N.G. 1990. Plant macrofossils. In: A.V. Morgan, ed. Short Course 2: Biological techniques in paleoenvironmental interpretation. First Joint Meeting, Canadian and American Quaternary Associations, Waterloo, Ontario. Pp. 30–72.

MILLER, N.G. 1992. A contribution toward a history of the Arctic moss flora. Contr. Univ. Michigan Herb. 18:73–86.

MILLER, N.G. 1993. New Late-Pleistocene moss assemblages from New England, U.S.A., and their bearing on the migrational history of the North American moss flora. J. Hattori Bot. Lab. 74:235–248.

NOWAK, C.L., R.S. NOWAK, R.J. TAUSCH, and P.E. WIGAND. 1994. A 30,000 year record of vegetation dynamics at a semi-arid locale in the Great Basin. J. Veg. Sci. 5:579–590.

SPAULDING, W.G., J.L. BETANCOURT, L.K. CROFT, and K.L. COLE. 1990. Packrat middens: their composition and methods of analysis. In: J.L. Betancourt, T.R. Van Devender, and P.S. Martin, eds. 1990. Packrat middens: the last 40,000 years of biotic change. University of Arizona Press, Tucson. Pp. 59–84.

SUGITA, S. 1994. Pollen representation of vegetation in Quaternary sediments: theory and method in patchy vegetation. J. Ecol. 82:881–897.

THOMPSON, R.S. 1988. Western North America. Vegetation dynamics in the western United States: modes of response to climatic fluctuations. In: B. Huntley and T. Webb III, eds. Vegetation history. Kluwer, Dordrecht. Pp. 415–458.

VAN DE WATER, P.K., S.W. LEAVITT, and J.L. BETANCOURT. 1994. Trends in stomatal density and $^{13}C/^{12}C$ ratios of *Pinus flexilis* needles during the last glacial-interglacial cycle. Science 264:239–243.

WATTS, W.A., B.C.S. HANSEN, and E.C. GRIMM. 1992. Camel Lake: A 40 000-yr history of vegetational and forest history from northwest Florida. Ecology 73:1056–1066.

WEBB, R.S., D.M. ANDERSON, and J.T. OVERPECK. 1994. Editorial: Archiving data at the World Data Center-A for Paleoclimatology. Paleoceanography 9:391–393.

WEBB, T. III. 1988. Eastern North America. In: B. Huntley and T. Webb III, eds. Vegetation History. Kluwer, Dordrecht. Pp. 385–414.

WEBB, T. III. 1993. Constructing the past from late-Quaternary pollen data: temporal resolution and a zoom lens space-time perspective. In: S.M. Kidwell and A.K. Behrensmeyer, eds. Paleontological Society Short Courses in Paleontology 6: Taphonomic approaches to time resolution in fossil assemblages. Paleontological Society. Pp. 79–101.

WEBB, T. III, P.J. BARTLEIN, S. HARRISON, and K.H. ANDERSON. 1993. Vegetation, lake levels, and climate in eastern North America for the past 18,000 years. In: H.E. Wright, Jr., J.E. Kutzbach, T. WEBB III, W.F. RUDDIMAN, F.A. STREET-PERROTT, and P.J. BARTLEIN, eds. Global climates since the last glacial maximum. University of Minnesota Press, Minneapolis. Pp. 415–467.

PROSPECTS FOR A DIGITAL FLORA OF NORTH AMERICA

JOHN T. KARTESZ

Biota of North America Program
North Carolina Botanical Garden
University of North Carolina
Chapel Hill, NC 27599, USA

ABSTRACT

Traditionally, textbook-style floras for previously "untreated" regions have often required many years to complete. They have usually included descriptions, keys, and illustrations, with some providing county-level distribution maps, and for large geographic areas they often have been produced as a series of separate volumes. Such serial floras have limited use until the final volume is completed, with the earlier volumes moving into various stages of obsolescence. With advances in computer technology, it is now possible to establish a more useful and efficient approach, by developing floristic treatments as a series of layered, linkable, digital databases. Each database layer can be maintained, updated, expanded, or published separately, and all can be interlinked to provide full coverage of the flora. Such information systems also are searchable in various ways. Recent advances in storage technology have produced various media that can contain more than four gigabytes of digital information, more than enough to provide the accepted names and synonyms for all taxa in the North American flora, their state and county-level distribution maps, including citations and vouchers, along with text descriptions, images, and digital keys.

RESUMEN

Tradicionalmente, las floras, en estilo libro, de regiones previamente "no tratadas" han requerido frecuentemente muchos años para completarse. Normalmente han incluido descripciones, claves e ilustraciones, con algunos mapas de distribución a nivel regional, y cuando las áreas son grandes se ha producido una serie de volúmenes separados. Estas floras seriales han tenido un uso limitado hasta que sale el ultimo volumen, mientras que los iniciales se van quedando obsoletos. Con los avances en informática, ahora es posible establecer una aproximación más útil y eficiente, desarrollando tratamientos florísticos como una serie de bases de datos digitales interrelacionadas. Cada parte de la base de datos puede mantenerse, actualizarse, expandirse, o publicarse independientemente, y todas ellas pueden interconectarse para dar una cobertura total de la flora. Estos sistemas de información son explorables de varios modos. Los avances recientes en la tecnología del almacenamiento han producido varios medios que pueden contener más de cuatro gigabytes de información digital, más de lo necesario para contener los nombre válidos y sinónimos de todos los taxa de la flora norteamericana, los mapas de distribución a nivel de estado y condado, incluyendo citas y testigos, junto con descripciones, imágenes y claves digitales.

TRADITIONAL NORTH AMERICAN FLORISTICS

For nearly 400 years, North American botanists have published traditional textbook-fashioned descriptive floras designed primarily to assess the various kinds of plants found growing somewhere on the continent. Normally written by one or two individuals, these floras have routinely included morphological descriptions, dichotomous keys, and plant illustrations, with some more enthusiastic state-level and regional workers providing county-level distribution maps within their floras as well. The average amount of time required to prepare a descriptive flora has historically been scale-dependent. Most US state-level floras have required 8 to 20 years to complete. E.G. Voss's *Michigan flora* (1972–1996), which began in 1954 and recently concluded in 1996, is an obvious exception. Regional floras have traditionally required even more time, ranging from 20 to 30 years to complete. Here, too, obvious exceptions exist, i.e., Cronquist et al.'s *Intermountain flora: vascular plants of the Intermountain West* (1972–1997), which began in the mid-1960's as a basic checklist and has now progressed through six published volumes and will likely require an additional six or more years to complete.

Although as a community we have not yet completed a textbook-fashioned flora for the North American continent, several international floristic efforts have been attempted. For example, the *North American flora*, published through the New York Botanical Garden, began in the early 1900's as a series of periodically published volumes, intended to describe all plants found growing on the continent. It continued through 1984, with the publication of the Linaceae (Rogers, C.M. 1984. *North American flora. Series II, part* 12), but the flora itself remains incomplete. *Flora North America* (Flora of North America Editorial Committee, Eds. 1993+), which actually began in the 1960's, with numerous starts and stops along the way, has similar intentions but on a slightly reduced geographic scale. It is currently slated to include some 30 volumes (three of which have been published), which will likely require decades to complete! Aside from the enormous time required to complete these serial floras, not to mention the high development costs and excessive paper consumption, another obvious consequence, whether state, regional, or international in scope, is that until the final volume appears, the flora as a whole has only limited value. Equally consequential is that by the time the final volume is published, the previous volumes may be obsolete. This is especially true for floras that languish over several decades before completion. Although serial textbook-fashioned floras have long been established as the paradigm for floristic writing in North America and elsewhere, given the advances in modern technology, it is now possible to break free from that tradition, perhaps to establish a new paradigm for floristic research in the 21st century.

ADVANCES IN TECHNOLOGY AND FLORISTIC UNDERSTANDING

Advances in personal computers, database technology, programming languages, Internet capabilities, and various novel digital products have allowed the botanical community to process floristic data with much greater speed and efficiency than ever before, providing us good prospects for producing a continental flora. Unlike floristicians in the 19th and early 20th centuries, today's workers can take full advantage of many of the very useful tools that were unavailable to them even a decade ago. For example, numerous international authority files and indices are currently available, offering assistance with subjects such as standardizing author abbreviations, providing photographs of specimens (including type material) for plants growing in many areas of the country, offering bibliographic reference guides, and depicting phytogeographic atlases and range data. It is now possible to access much of the necessary nomenclatural and taxonomic literature needed to build a flora while sitting at an Internet-connected computer terminal. Although obviously not all of the computer-related problems have been resolved regarding floristic research, computers have greatly enhanced the manner in which floras can now be developed. By taking full advantage of all of the computer technology currently available and being open to a new paradigm for floristic research, we can now establish a new methodology for producing floristic references in a far more efficient manner than ever before.

Simultaneous with the advances in technology have come enormous advances in our understanding of North American floristics. Indeed, it is entirely reasonable to conclude that relatively few mysteries remain regarding our knowledge of the North America flora. That is not to say that all of our plants have been described, and that the taxonomy of all genera including *Rubus* and *Crataegus* has been completely assessed, nor is it to say that we know every state-level occurrence for every taxon. We are, however, extremely knowledgeable about the number and types of plants that occur throughout the continent (or at least for that area north of Mexico), and we are equally knowledgeable about the plant distributions within this area. Over the past decade or so, it has become common knowledge that there are about 22,000 species found in North American north of Mexico (understanding fully that different taxonomies will always produce a slightly different total). We are equally knowledgeable that some 18,000 native species occur within this region, that the number of species introduced from foreign areas is nearly 4,000, that the total number of state-level endemic taxa is nearly 6,000; and that there are nearly 2,400 tree species, 4,000 shrub species, 17,000 herbaceous species, 900 fern species, 2,600 grass species, etc. We also know that nationally, there are nearly 20,000 species in the US and that the average number of species found within each US state is approximately 2,700; that there are approximately 5,300 species in Canada; that there are approximately 600 species in Greenland; and that regionally, there are some 7,400 species in the Southeast, 6,000 species in the Northeast, 7,800 species in the Great Plains, etc. (Kartesz & Meacham 1999).

A NEW FLORISTIC PARADIGM

The recently completed *Synthesis of the North America flora* (Kartesz & Meacham 1999) provides these and countless other summaries as well as compelling evidence that a reasonable alternative now exists to the classic serial flora. This publication offers a new paradigm for floristic development and provides an approach that takes full advantage of many of the recent advances in computer technology. It also offers an approach to produce a fully dynamic and functional flora for North America based primarily upon numerous, individually-layered floristic databases, all linked together beginning with the most fundamental layer, a system of nomenclature and taxonomy. This is followed by the second layer, a phytogeographic atlas, which in turn is followed by numerous layered databases for biological attribute data, with the possibilities of illustrated polyclaves and plant descriptions to follow. These layered databases are linked together via an exceptionally fast search engine that provides nearly instantaneous response time for even the most complex queries. Now completed, the various layered databases, which are internal, can be further linked to other external databases, including illustrated random access identification keys, photographs, descriptions, and even alternative taxonomies. There is virtually no end to the number and kinds of linked data fields that can be added to the system. Indeed, one of the more obvious benefits of this approach over the serial floras is that the various individual floristic databases can be developed rather quickly, easily updated, and published independently. Moreover, each layer can be queried individually or in Boolean fashion with other layers, enabling wholly unique subsets of data such as county, state, regional, or even national checklists, atlases, keys, biological summaries, etc., to be produced as fully functional subunits of the total flora. This system also enables checklists to be developed for any area of the continent by offering the names that are necessary to produce floristic summaries for geographic areas ranging in size from a backyard or private woodlot to a US state or region.

Once linked together, layered floristic databases become immeasurably more useful than a static and rigid serial flora. For example, in the southeastern US it is possible to establish immediately the number and type of species present within the region or within each state, or the number of native species, or the number of introductions, etc. Indeed, with the various layered databases all linked together, it is difficult to calculate the actual number of data combinations possible for the Southeast, but certainly, the number would be astronomical. Conversely, with a serial flora, even when completed, the utility would be limited such that few if any meaningful comparisons would be possible, and even the few that were, would require substantial time and effort. I have provided a table to illustrate the types of floristic data summaries that are possible for the 14 southeastern states using the *Synthesis of the North American Flora* (see Table 1).

UPGRADING LAYERED DATABASES

The three most fundamental components of any flora, including the layered database system described above are (1) the taxonomy, (2) the nomenclature, and (3) the phytogeography. Aside from being the most fundamentally important, they are also the most variable and thus require the most effort to update. It is therefore essential that in producing this type of floristic system, functional and effective data bridges be established to enable the linking of these three databases with others within the system. A data bridge is a computer program that automatically links or disengages certain data fields from other data fields. When programmed efficiently, data bridges can save enormous amounts of time and effort; however, with complex data fields such as those necessary for developing a continental flora, data bridges, no matter how efficiently constructed, will not accommodate all the necessary data links and/or disengagements. Therefore, it must be expected that some of the links and disengagements will require manual application.

SPECIALIZED LAYERED DATABASES FOR VOUCHERING AND KEYING

One of the more specialized types of floristic data that has gained interest over the recent past is vouchering data. Vouchering of phytogeographic records has historically been limited primarily to monographic and revisionary publications. Very few historical floristic works in the US, with the exception of C.C. Deam's *Flora of Indiana* (1940) and a few others, have provided voucher data for plant occurrences. Thus, suspicion and speculation abound regarding literature reports of geographic distribution records found in many of our regional and state floras. Even the most recent state-level and regional floras tend to provide little to no documentation for occurrence

TABLE 1. A floristic summary of the southeastern states.

	AL	AR	DE	FL	GA	KY	LA	MD	MS	NC	SC	TN	VA	WV
Total sp.	3550	2697	2118	4121	3630	2676	3122	3104	2854	3660	3354	2984	3419	2409
Native sp.	2946	2227	1676	3120	3057	2096	2442	2318	2396	2913	2715	2427	2662	1899
Introduced sp.	604	470	442	1001	573	580	680	786	458	747	639	557	757	510
% Introduced	17.0	17.4	20.9	24.3	15.8	21.7	21.8	25.3	16.0	20.4	19.1	18.7	22.1	21.2
Endemic taxa (excl.hybs.)	19	11	1	261	27	4	3	1	1	9	2	12	14	11

data. This has resulted in numerous errant records, which are especially evident for those of us in the Southeast who have from time to time relied on J.K. Small's publication *The manual of the southeastern flora* (1933), where countless erroneous state-level attributions were made. According to W.H. Camp (Castanea 16(3): 115. 1951.):

Occasionally Small indicates the presence of a plant in some area for which there is no known authenticating specimen. I once asked Small about some of these and he grinned and admitted that, in certain instances, the plants had not yet been collected but that, sooner or later they would be–that he [Small] had put those in just to make botanists "dig" a little harder.

Regrettably, many of these erroneous reports have been perpetuated in both state and regional floras.

Over the past decade, an increased demand has been placed on floristic efforts to include voucher data authenticating plant occurrence data at both state and county levels. Citing voucher data for all taxa within a separate database, for every geographic area, even if it were simply an acronym for the repository housing a particular specimen, would be enormously useful for state or county level floristic documentation. Given the remarkable increases in both memory capacity and processor speed in today's computer systems, this would not be a difficult task, at least for the various computer hardware and software systems.

In recent years, a number of universities and museums have attempted to computerize their holdings, for the purpose of making specimen data available to others via the Internet. There has been at least some debate over the usefulness of this effort, even though prestigious institutions such as the National Science Foundation have funded some institutions to work toward accomplishing this goal. In some cases, in addition to providing label data for each specimen, the documentation process has included images of specimens as well, which have also been made accessible via the Internet. It has been estimated that some 70 million specimens are now housed in our North American repositories, which makes the task of entering all of the label data for this number of specimens very difficult, if not impossible. For small collections (those with fewer than 100,000 specimens) the task is certainly manageable. However, for large collections (with 500,000 or more holdings) the task has proven to be very demanding, very expensive, and perhaps of questionable scientific merit. Although it would be somewhat useful to have all specimens vouchered and accessible in this way, the resources necessary to accomplish the task may be prohibitive. Moreover, the benefits do not appear to be immediately obvious. For example, to understand that several hundred specimens exist for a particular taxon within a specified locality, say a county of an eastern state, and to provide the necessary label data to document this, offers very limited ecological, evolutionary, systematic, or even phytogeographic benefit, especially knowing that so many of our specimens are represented by non-random collections, i.e., most collections are made from areas along highways, roadways, or other right-of-ways, near universities, or otherwise gathered from rather popular collection sites.

Perhaps a more pragmatic solution to the problem would be to establish a single *bona fide* voucher for every taxon known to occur within a particular US state or county or equivalent area of the continent. Even using this approach, it would likely require nearly 500,000 vouchers to be located and assessed at the state level, and 5 million voucher specimens to be located and assessed at the county level.

Nonetheless, an individual-vouchered database prepared in this manner and linked to other floristic data would add significantly to the credibility of our national or international phytogeographic atlases. Admittedly, employing this approach would provide only minimal documentation, but it would solve at least one problem, i.e., provision of reliable evidence for the occurrence of a particular taxon within a particular geographic area.

A second specialized component or layer that will likely accompany future floristic research in North America, as well as elsewhere, is a polyclave (also known as a random-access identification key). A polyclave ("many key") is essentially a taxon/character data matrix that permits unrestricted choices of characters and character states, beginning with whichever ones are selected by the user as being most conspicuous or most diagnostic. Once completed, a polyclave could permit the user to identify the members of any floristic group, with a high degree of reliability, using the most conspicuous, most easily understood, and/or most easily defined plant characters and character states, such as flower color, number of petals, leaf type, leaf position, etc. With relatively few choices, a step-wise elimination quickly narrows the taxonomic possibilities. It could be possible to advance this concept further by incorporating illustrations to create a "visual polyclave," whereby at any point in the selection process, the user could display a group of images (representing those species not yet eliminated) and a tentative identification could be made by matching the unknown plant at hand with one of the images on the computer screen. It is interesting to note that up to 60 or more illustrations could be loaded onto a single computer screen. Any of these illustrations could be enlarged and the accompanying morphology could be presented. In most cases, with only a few initial choices of geography and easily observable morphology and a choice from a display of images, an unequivocal identification could be made. A visual polyclave, developed as separate layer of a digital floristic work, could certainly function as a stand-alone product, but could also be designed to be integrated into other database layers as well.

Over the past decade, polyclave research and theory has become far more advanced, challenging computer programmers to produce software that meets the needs of the biologists. Polyclave identification clearly represents a trend that cannot be ignored. It has now achieved world-wide interest in many areas of biology and researchers in many fields are building digital keys that offer much more ease and versatility than do traditional dichotomous keys. Once completed, illustrated polyclaves could represent the quickest and perhaps easiest way to achieve species level identifications.

For the North American flora, several centers have made significant progress in polyclave research, including XID Services, Inc., the New York State Museum, and the Biota of North America Program (BONAP) at the University of North Carolina. The entire vascular flora at both the family and generic levels has now been completed by BONAP staff in draft polyclave format, and nearly half of the North American species have also been completed using this approach.

INTERNET APPLICATION TO THE NEW FLORISTIC PARADIGM

Now that the Internet has gained nearly universal acceptance and global use, the next logical step in North American floristics is to make floristic layered databases available to the general public. There could be tremendous advantages for doing so. For example, providing these types of databases on the Internet and circulating them to a national and international community could greatly enhance the interaction between those responsible for developing the layered databases with monographers, floristicians, and others within the community who might be willing to provide feedback to keep the databases current. The layered databases could also be updated frequently and new editions could be posted to the Internet, without the tremendous time delay often associated with traditional hard-copy publications or even with CD-ROM or diskette upgrades for software designed for use on stand-alone computers. Equally advantageous could be the low cost of connectivity, especially for students and faculty in colleges, universities, and other public and private institutions that promote educational activities. Through the Internet, it could also be possible to link data from one Web site with another, and to access various related data, including plant illustrations, photographs, descriptions, etc.

Unfortunately, there could be some obvious drawbacks with using the Internet. The current slow speed of the Internet, at least in some locations, might inhibit scientific research that could proceed more quickly with stand-alone computers or with traditional reference works. This is especially true in rural communities, where slow and inefficient Internet connections abound. For those who might have to purchase their Internet access, this could also become somewhat expensive. Until wireless technologies are dramatically improved, it would be impossible to rely on the Internet for conducting field research; thus floristic data sets posted on the Internet would likely continue to be used nearly exclusively in herbaria, libraries, homes, and classrooms. Moreover, due to limitations in speed and the complexity of the programming involved, it will remain difficult to run exceptionally complex database searches or other queries via the Internet. Exacerbating the problem further would be that data designed for release on the Internet often lack critical archival documentation, making them very risky to cite. It is often the case that Web site documents in general, whether they be state rare plant listings, summaries of state or county-level floras, or collection data are altered or otherwise updated periodically, depending upon the level of research activity by the publishing author. In some cases, such data sets are very short-lived and quickly and unexpectedly withdrawn. It is obviously very difficult to rely on such data. An additional but related problem would be that through attempts made by some groups or individuals to purchase or otherwise obtain large data sets and thereafter incorporate them into their own Internet sites, some data centers would need to alter substantially the data contained within the acquired layered databases for the purpose of making them compatible with their own systems. In doing so, they would occasionally and often unwittingly change the intent of the original data. These types of problems would need to be corrected in order to establish and maintain scientific credibility.

The much greater speed-of-access achievable by stand-alone systems, such as desktop, laptop, or palmtop computers, and the ease by which these systems can be programmed and adapted for field use (at least for palmtop and laptop computers), strongly suggests that the future of floristic field and laboratory references lies in these types of digital products. The speed by which data can be accessed via this medium would depend only upon the quality of the user's computer hardware and the quality of the software program that has been written to provide data access. With the continual advances in program languages and tool development, which makes it much easier for programmers to write more complicated applications, there is virtually no limit to the types of queries that can be run. Once purchased or downloaded from the Internet, these programs that run on free-standing computers are readily accessible to the users and can be transported and used in the field or laboratory. Another advantage of free-standing computers is that programs written for these machines also include critical archival documentation, such as the author's name, title and date of publication, enabling the data to be cited accurately.

Although software programs designed for free-standing computers are not plagued with the number or types of problems of the Internet, they, too have limitations and shortcomings. For example, since software programs incorporate the most advanced programming languages, users are often forced to continually upgrade their computer hardware if they want to employ the latest software. Fortunately, as the quality of computer hardware increases, prices will continue to fall, so perhaps this will no longer be viewed as the type of hardship that it had been in the past. Nonetheless, annual or periodic upgrades to these software programs will be necessary for the user to continue to have the most advanced data. Also, since software programs for free-standing machines will not be circulated as widely as the data on the Internet, the authors of these data will likely receive fewer comments from members of the botanical community.

Perhaps the best solution in preparing a continental digital flora would be to incorporate the advantages offered by both the Internet and the free-standing computer. In combination, it will be possible to engage the largest membership of the national and international communities via the Internet, while maintaining the speed, ease of programming, and potential for field and laboratory use of the data on laptop or palmtop computers. By employing both systems, it would also be possible to produce an annual, archival upgrade of the flora on a CD-ROM or a series of diskettes for use on stand-alone computers, while continually providing regular upgrades to the system on a related Web site. This approach would keep the system current and yet provide the most flexibility and ease-of-use.

Regardless of what medium is used, any continental flora for North America will only be truly successful if it receives community-wide involvement and support. The earlier this is accomplished and the easier it is for the community to participate, the more likely it will be successful.

ACKNOWLEDGMENTS

I thank Guy L. Nesom and Amy C. Farstad for helpful comments.

REFERENCES

CAMP, W.H.. 1951. On the proposed checklist of southeastern plants. Castanea 16:112–115.

CRONQUIST, A. ET AL. 1972-1997. Intermountain flora: Vascular plants of the Intermountain West. Columbia University Press, New York.

DEAM, C.C. 1940. Flora of Indiana, plus unpublished and undated additions by L. Johnston. Wm.B. Burford Printing Co., Indianapolis.

FLORA OF NORTH AMERICA EDITORIAL COMMITTEE, EDS. 1993+. Flora of North America north of Mexico. 3+ vols. Oxford University Press: New York and Oxford.

SMALL, J.K. 1933. The manual of the southeastern flora. Published by the author, New York.

KARTESZ, J.T. 1999. In: Kartesz, J.T. and C.A. Meacham. Synthesis of the North American flora (Ver. 1.1). Published by the Biota of North America Program of the North Carolina Botanical Garden, Chapel Hill.

ROGERS, C.M. 1984. Linaceae. North American flora. Series II, pt. 12. New York Botanical Garden, Bronx.

VOSS, E.G. 1972–1996. Michigan flora, volumes 1–3. Cranbrook Institute of Science and University of Michigan Herbarium, Bull. Nos. 59–61.

APPLICATION OF DELTA IN FLORISTIC BOTANY

ROBERT D. WEBSTER

United States Department of Agriculture
Agriculture Research Service, Plant Sciences Institute
Bldg. 003, Rm. 106
Beltsville, MD 20704, U.S.A.

ABSTRACT

DELTA (DEscriptive Language for TAxonomy) is a validated system specifically designed to effectively manage data associated with a floristic treatment. The system offers an effective mechanism for the initial recording, updating, checking, storage, and retrieval of floristic data. Natural language descriptions as generated by DELTA, have the potential to be technically superior to the conventional floristic description. These descriptions are consistent in character sequence, punctuation, and can include the best or most appropriate subset of characters. DELTA generated identification keys allows the flora writer to superimpose his concepts of character reliability and key design on the programs requirements for correctness and efficiency. Floristic data in DELTA format converted to the interactive format allows for identification procedures and information retrieval aspects superior to those of the traditional flora. Data recorded in DELTA format is available to future advances in computer technology.

RESUMEN

DELTA (DEscriptive Language for TAxonomy) es un sistema diseñado específicamente para manejar de modo efectivo datos asociados con un tratamiento florístico. El sistema ofrece un mecanismo efectivo para el registro inicial, puesta al día, comprobación, almacenamiento y recuperación de datos florísticos. Las descripciones en Lenguaje Natural como son generadas por DELTA, tienen el potencial de ser técnicamente superiores a las descripciones florísticas convencionales. Estas descripciones son constantes en la secuencia de caracteres, puntuación, y pueden incluir el mejor o más apropiado subconjunto de caracteres. Las claves generadas por DELTA permiten al autor de la flora superponer su concepto de confianza en el carácter y el diseño de la clave con los requisitos del programa para la corrección y eficiencia. Los datos florísticos en el formato DELTA convertidos al formato interactivo permiten procedimientos de identificación y recuperación de información superiores a los de la flora tradicional. Los datos grabados en formato DELTA están disponibles para los futuros avances en la tecnología informática.

INTRODUCTION

Description Language for Taxonomy (DELTA) is the internationally accepted standard for encoding taxonomic descriptions (International Taxonomic Databases Working Group). World-wide web access to all relevant information of the DELTA system can be found at biodiversity.uno.edu/delta. Data in DELTA format (Dallwitz, 1980; Dallwitz and Paine, 1993 onwards) is linked in the system to the production of descriptions, key generation, and information retrieval systems. The initial validation of the technology was made by Watson et al. (1971–1992) in his work with the legume genera, grass genera of the world, and plant families of the world. These are monumental pieces of research that are recognized worldwide as standards in the application of computer technology to taxonomic monographs. Early in the 1980s Webster made the first application of the system to a critical taxonomic revision. That was followed by application of DELTA to a floristic treatment of the Panicoid grasses of Australia. From 1985 to 1995, Webster published numerous taxonomic treatments (mainly in SIDA) where detailed discussions on procedures were given. Aiken (1991–1997) contributed additional work with grasses. Argus (1991, 1995, 1997) effectively applied the system in treatments of *Salix.* Bruhl (1992–1995) made an outstanding contribution with his treatment of the Cyperaceae. Kirkbride (1993–1994) made innovative applications in his critical research in *Cucumis.* O'Brien (1992–1994) made a very effective application of the technology in his work on the Coleoptera. Additional references to the DELTA system and the application of the DELTA system are given here. A comprehensive list of references is available at the DELTA web site. The objective of this paper is to provide an introduction to the system as related to the production of a flora.

SIDA, Bot. Misc. 18: 129–135. 2000

THE DATA

The most important aspect of applying the DELTA technologies to a floristic account is the initial organization of the floristic data. However, as any project progresses changes and improvements can be made in these basic files. Fundamentally, this refers to two basic data resources; the characters and the recorded data for the taxa. The character files represents an organized sequence of independent mutually exclusive characters. Examples of characters include spikelet shape, spikelet length, spikelet width, spikelet color, shape of an apex, shape of a base, etc. Each character is associated with the mutually exclusive character states applicable to the study. It is a good idea to define the characters and character states as they are constructed. The character file may include numerous characters. Even a small study of a relatively limited number of taxa may require a character list of 100 characters. For floristic accounts, the character file may well consist of hundreds of characters. Construction of the character file is of critical importance to the success of a project. However, there are procedures within the system to update, modify, and change the character file as the project evolves.

The second fundamental data file is the so-called items file. This file contains the taxa with data recorded in DELTA format. That is, an organized sequence of taxon names followed by the recorded data for the characters. Included in the items file is miscellaneous comment data and additional text information. Organization of the items file is determined by the database developer. All taxa may be maintained in one file or in separate files. For example, families may be maintained in a file separate from genera and separate from sub-generic categories. DELTA includes flexibility to allow the developer to organize the system to meet the specific requirements of the study. Data in the items is recorded in free field format, where there are no restrictions on the beginning or ending of the file. Data may be entered in any order. The system has been designed for the convenience of taxonomists. An appreciation of the system is best understood through a practical application.

The remaining files associated with a typical DELTA directory are associated with database management. The most important of these include files to automatically check the data files, files to reorder the data, and files to automatically print specific output. The data files can be automatically converted to natural language description, a format for traditional key production, interactive INTKEY format, DIST format, PAUP format, Nexus format, Henning format, ALICE format, and NT-SYS format. The conversions to descriptions, key format, and interactive format are directly related to the production of a flora.

DESCRIPTIONS

The DELTA System converts recorded data into technically superior natural-language descriptions. The system guarantees order and consistency in character sequence and in the use of punctuation. This results in easy to interpret highly comparable descriptions which can be designed to meet the specific needs of a flora. Whereas data may be available for many characters, the description that appears in print can be restricted to a subset of the data. For example, a description limited to vegetative characters, inflorescence characters, or ecological characters can be used. DELTA allows for the production of a description that meets specific editorial restrictions in terms of format and length. The following description of *Axonopus* P. Beauv., published by Webster et al. (1989), illustrates some basic advantages offered by the system.

Axonopus P. Beauv.

Internodes solid or spongy. Ligule a ciliate membrane or a fringe of hairs. Main axis with quaquaversal primary branches. Primary branches with appressed secondary branches; with secund spikelets. Rachis terminating in a spikelet. *Pedicels concave at the apex. Disarticulation at the spikelet base. Spikelets adaxial*; dorsiventrally compressed or planoconvex. *First glume absent*. Second glume subequal to spikelet length. *Lemma of upper floret cartilaginous*; smooth or muricate; with flat of involute margins; muticous.

In the above description, a diagnostic subset of the recorded data is presented. For each recognized genus approximately 300 characters were recorded. An analysis was preformed to select the diagnostic subset of characters for each taxon. Implied characters were automatically omitted from the descriptions. For example, the statement "first glume fused with the callus to form a cup-like structure" would be given for *Eriochloa*, but the implied negative is not repeated for the remaining genera. A list of the diagnostic characters was generated. This list was used to produce the published floristic descriptions. In addition, the diagnostic characters for each genus is italicized within each description. In the above description, the italicized characters distinguish the genus from other taxa in the study. The results of this procedure is a technically superior description ideal for a floristic treatment. With data recorded in DELTA format, it is relatively easy to produce descriptions in different languages. This is accomplished by converting the character list to the desired language. The accompanying microfiche in Webster et al. (1989) included comprehensive taxonomic treatments in both Portuguese and Spanish. In summary, the DELTA system offers the potential to automatically generate technically superior floristic descriptions.

IDENTIFICATION KEYS

Data in DELTA is available for direct conversion to KEY format for the production of both traditional bracketed and indented keys. The initial advantage of the system is the agreement between the descriptions and the diagnosis with the keys. That is, since the data producing the descriptions is the same as the data producing the keys, there will be total agreement between the two, a relationship not always present in traditional products. The system allows for the user to quickly experiment with key designs That is Key allows the user to superimpose and vary his concepts of character reliability on the programs analysis for mechanically efficiency. The conversion of recorded data for key-making purposes requires two important subjective processes. These include the assigning of character weights and the defining of key states for multistate and numeric characters. If one desires a strictly dichotomous key then each character must be defined in terms of two states. For example, spikelet length or length of a petal is best recorded as a real numeric character. That is, spikelets 3.2–3.7 mm long or petals 5.7–7.8 mm long. This character for key making purposes is redefined under two states. For example, less than 4.0 mm and 4.0 or greater. Recognition of the two states is subjective and determined by the distribution of the data or designed to make a specific distinction. Assigning of character weights is the second important aspect to be considered. Each character can be assigned a relative weight between 0 and 9, with 5 the default, and nine the most important. With the defining of the key states and assignment of character weights a key can be generated which reflects the parameters and distribution of the data. The output key will be technically correct and the perceived value or usefulness of the key is determined by the opinion of the producer or user. Generally, a user of this system with an opinion of key design will use the system as an experimental process. That is, numerous keys are generated and parameters are varied until a final product is produced that meets the systems requirements of correctness and efficiency and the taxonomists concepts of character value and key design. The use of the key making system is best applied to large problems. That is, if a key to 4 taxa is needed, the system would be of limited value. In other words, it is very easy to hand produce a key to a few taxa. However, if a key to 20, 50, or 200 taxa is required then the DELTA system becomes an invaluable tool for experimentation in key design. Once a satisfactory key is produced, there may be a subjective need to make editorial changes. Such changes should be made with care since it is easy to incorporate a concept not in the original data. In addition, the sequence of secondary characters within a couplet should not be edited to meet editorial requirements of saving space. In conclusion, the DELTA system offers the user the capability of producing technically superior keys that are mechanically correct. In addition, it allows the taxonomist to superimpose their concepts in the key design.

INTERACTIVE KEYS AND INFORMATION RETRIEVAL

Conversion of the data to the interactive format provides numerous important advantages over the conventional floristic products of descriptions and keys. The interactive procedure was designed for identification, information retrieval and comparison, and database interrogation. In the identification process, the most immediate advantage is the ability of the user to use any character at any point in the identification process. With conventional keys, users are restricted to the characters selected by the author. With interactive keys, identification can be based on user selected characters. In addition, the system can be set to allow an error in the identification process. Interactive keys allow for the use of unclassified real data. That is, in conventional keys numeric data is classified into categories to produce the desired dichotomy. For example, spikelets less than 4.0 mm and 4.0 mm or greater. Interactive keys allow for the identification based on real data. A measurement may be taken from a specimen, entered and the taxa with that specified range then displayed. The interactive identification process will suggest the best characters for identification. Selection of these "suggested best characters" would be influenced by the user concepts of the reliability of the characters.

An especially important aspect of the system is the ability to link the characters and taxa to illustrations. If there is a question concerning the meaning of a character, an illustration can be displayed that explains the character. Likewise, an illustration of a taxon may be linked to the data. Information on the similarities and differences among taxa can be immediately displayed. Data stored in text format may be searched and displayed. The interactive procedure allows for the analysis of diagnostic characters and for the inclusion of notes or glossaries on the characters and taxa. Information retrieval with the system may be best explained by example. The system would provide an immediate answer for the taxa from a defined geographic region, with a specified morphology, and phenology. For example, the annual taxa from Louisiana with spikelets 3.2-3.7 mm long, first glume 3-nerved, flowering in June, and diploid. Conventional floras do not provide for this type of important information retrieval. Finally, the system efficiently treats very large datasets in terms of characters and taxa. In conclusion, computer technology illustrated by the interactive identification and information retrieval process mandates that modern floristic accounts include these technologies.

SUMMARY

DELTA (DEscriptive Language for TAxonomy) is a validated system specifically designed to effectively manage data associated with a floristic treatment. The system offers an effective mechanism for the initial recording, updating, checking, storage, and retrieval of floristic data. Natural language descriptions as generated by DELTA, have the potential to be technically superior to the conventional floristic description. These descriptions are consistent in character sequence, punctuation, and can include the best or most appropriate subset of characters. DELTA generated identification keys allows the flora writer to superimpose his concepts of character reliability and key design on the programs requirements for correctness and efficiency. Floristic data in DELTA format converted to the interactive format allows for identification procedures and information retrieval aspects superior to those of the traditional flora. Data recorded in DELTA format is available to future advances in computer technology.

REFERENCES

AIKEN, S.G., L.L. CONSAUL, and M.J. DALLWITZ. 1996. Grasses of the Canadian Arctic Archipelago: A DELTA database for interactive identification and information retrieval. Canad. J. Bot. 74:1812–1825.

AIKEN, S.G. and M.J. DALLWITZ. 1991. *Festuca* (Poaceae) of North America: Interactive identification and information retrieval. Flora Online 26.

AIKEN, S.G., M.J. DALLWITZ, C.L. MCJANNET, and L.L. CONSAUL. 1997. Biodiversity among *Festuca* (Poaceae) in North America: Diagnostic evidence from DELTA and clustering programs, and an INTKEY package for interactive, illustrated identification and information retrieval. Canad. J. Bot. 75, 1527–1555.

AIKEN, S.G., M.J. DALLWITZ, C.L. MCJANNET, and L.L. CONSAUL. 1997. Fescue grasses of North America: Interactive identification and information retrieval. CD-ROM. Canadian Museum of Nature, Ottawa.

ARGUS, G.W. 1991. Salicaceae. In: G.W. Douglas, G.B. Straley, and D. Meidinger, eds. The vascular plants of British Columbia. Part 3. British Columbia Ministry of Forests, Special Report 3. Pp. 55–67.

ARGUS, G.W. 1995. Salicaceae: Willow family. Part two: *Salix*. J. Arizona-Nevada Acad. Sci. 29:39–62.

ARGUS, G.W. 1997. Infrageneric classification of *Salix* (Salicaceae) in the New World. Syst. Bot. Monogr. 52:1–121.

ASKEVOLD, I. S. and C.W. O'BRIEN. 1994. DELTA, an invaluable computer program for generation of taxonomic monographs. Ann. Entomol. Soc. Amer. 87:1–16.

ASKEVOLD, I.S., C.W. O'BRIEN, and K. MORIMOTO. 1994. Systematics and evolution of weevils of the genus *Bagous* Germar (Coleoptera: Curculionidae) III. Phylogenetic analysis of the species known from Japan. Esakia 34:75–129.

BOSWELL, K. F. and A.J. GIBBS. 1986. The VIDE (Virus Identification Data Exchange) project: A data bank for plant viruses. In: R.A.C. Jones and L. Torrance, eds. Development and applications in virus testing. (Association of Applied Biologists, UK. Pp. 283–287.

BRITTON, E.B. 1986. A revision of the Australian chafers (Coleoptera: Scarabaeidea: Melolonthinae). Vol. 4. Tribe Liparetrini, genus Colpochila. Aust. J. Zool., Suppl. Ser. 118:1–135.

BRUHL, J.J., L. WATSON, and M.J. DALLWITZ. 1992. Genera of Cyperaceae: Interactive identification and information retrieval. Taxon 41:225–235.

BRUHL, J.J. 1995. Sedge genera of the world: Relationships and a new classification of the Cyperaceae. Austral. Syst. Bot. 8:125–305.

BUECHEN-OSMOND, C. and M.J. DALLWITZ 1996. Towards a universal virus database—progress in ICTVdB. Arch. Virol. 141:392–399.

CALDER, A.A., J.F. LAWRENCE, and J.W.H. TRUEMAN. 1993. *Austrelater*, gen. nov. (Coleoptera: Elateridae), with a description of the larva and comments on elaterid relationships. Invertebr. Taxon. 7:1349–94.

CAMPBELl, C.S. and E.A. KELLOGG. 1987. Sister group relationships of the Poaceae. In: T.R. Soderstrom, K.W. Hilu, C.S. Campbell, and M.E. Barkworth, eds. Grass systematics and evolution. Smithsonian Institution, Washington. Pp. 217–224.

DALLWITZ, M.J. 1974. A flexible computer program for generating identification keys. Syst. Zool. 23:50–57.

DALLWITZ, M.J. 1980. A general system for coding taxonomic descriptions. Taxon 29:41–46.

DALLWITZ, M.J. 1992. A comparison of matrix-based taxonomic identification systems with rule-based systems. In: F.L. Xiong, ed. Proceedings of IFAC workshop on expert systems in agriculture. International Academic Publishers, Beijing. Pp. 215–218.

DALLWITZ, M.J. 1992+. An introduction to computer images. TDWG Newsletter 7:1–3. Reprinted in Delta Newsl. 8:7–9. Latest version at http://biodiversity.uno.edu/delta/

DALLWITZ, M.J. 1993. DELTA and INTKEY. In: R. Fortuner, ed. Advances in computer methods for systematic biology: Artificial intelligence, databases, computer vision. The Johns Hopkins University Press, Baltimore, MD. Pp. 287–296.

DALLWITZ, M.J. 1994. Desirable attributes for interactive identification programs. http://biodiversity.uno.edu/archive/taxacom.html

DALLWITZ, M.J. and T.A. PAINE. 1993+. Definition of the DELTA format. http://biodiversity.uno.edu/delta/

DALLWITZ, M.J., T.A. PAINE, and E.J. ZURCHER. 1993+. User's guide to the DELTA system: A general system for processing taxonomic descriptions. 4th ed. http://biodiversity.uno.edu/delta/

DALLWITZ, M.J., T.A. PAINE, and E.J. ZURCHER. 1993+. New features for the DELTA system. http://biodiversity.uno.edu/delta/

DALLWITZ, M.J., T.A. PAINE, and E.J. ZURCHER. 1998. Interactive keys. In: P. Bridge, P. Jeffries, D.R. Morse, and P.R. Scott, eds. Information technology, plant pathology and biodiversity. CAB International, Wallingford. Pp. 201–212.

DALLWITZ, M.J. and E.J. ZURCHER. 1988. User's Guide to TYPSET: A Computer Typesetting Program. 2nd ed. CSIRO Austral. Div. Entomol. Rep. 18:1–25

DORDA, E., and J. FERNÁNDEZ CASAS. 1989. Estudios morfológicos en el género *Narcissus* L. Anatomía de hoja y escapo, III. *Fontqueria* 27:103–162.

EDGAR, E. 1986. *Poa* L. in New Zealand. New Zealand J. Bot. 24:425–502.

GIBBS RUSSELL, G.E., L. WATSON, M. KOEKEMOER, L. SMOOK, N.P. BARKER, H.M. ANDERSON, and M.J. DALLWITZ. 1990. Grasses of Southern Africa. An identification manual with keys, descriptions, classification and automated identification and information retrieval from computerized data. Mem. Bot. Surv. South Africa 58:1—437.

GUNN, C.R. 1991. Fruits and seeds of genera in the sufamily Caesalpinioideae (Fabaceae). U.S.DA Tech. Bull. 1755. Government Printing Office, Washington, DC.

HUNT, G.S, M.J. COLLOFF, M.J. DALLWITZ, and D.E. WALTER. 1998. Oribatid mites: An interactive key to Oribatid mites of Australia. CD-ROM. CSIRO Publishing, Melbourne.

KAKUDIDI, E.K.Z., M. LAZARIDES, and J.A. CARNAHAN. 1989. A revision of *Enneapogon* (Poaceae, Pappophoreae) in Australia. Austral. Syst. Bot. 1:325–353.

KELLOGG, E. A. and C.S. CAMPBELL. 1987. Phylogenetic analyses of the Gramineae. In: Grass systematics and evolution. T.R. Soderstrom, K.W. Hilu, C.S. Campbell, and M.E. Barkworth, eds. Smithsonian Institution, Washington, DC. Pp. 310–324.

KELLOGG, E.A. and L. WATSON. 1993. Phylogenetic studies of a large data set. I. Bambusoideae, Andropogonodae, and Pooideae (Gramineae). Bot. Rev. 59:273–343.

KIM, S.P. 1990. Taxonomic revision of the Australian species of *Homoneura*, *Trypetisoma* and allied genera (Diptera: Lauxaniidae). Ph.D. Thesis, Australian National University, Canberra.

KIRKBRIDE, J.H. 1993. Biosystematic monograph of the genus *Cucumis* (Cucurbitaceae). Parkway Publishers, Boone, NC.

KIRKBRIDE, J.H. 1994. Revision of *Cucumella* (Cucurbitaceae, Cucurbitoideae, Melothrieae, Cucumerinae). Brittonia 46:161–186.

LAWRENCE, J.F., A.M. HASTINGS, M.J. DALLWITZ, T.A. PAINE, and E.J. ZURCHER. 1995+. Larvae of the Elateriformia (Coleoptera), families and sub-families: Descriptions, illustrations, identification, and information retrieval. http://biodiversity.uno.edu/delta/

LIEDE, S. 1996. A revision of *Cynanchum* (Asclepiadaceae) in Africa. *Ann. Missouri Bot. Gard.* 83:283–345.

LINDER, H.P. and R.P. ELLIS. 1990. A revision of *Pentaschistis* (Arundineae: Poaceae). Contr. Bolus Herb. 12:1–124.

LOWRY, J.K. and H.E. STODDART. 1993b. The *Onisimus* problem (Amphipoda, Lysianassoidea, Uristidae). Zool. Scripta 22:167–181.

MACFARLANE, T.D. 1979. A taxonomic study of the pooid grasses. Ph.D. Thesis, Australian National University, Canberra.

MACFARLANE, T.D. 1987. Poaceae subfamily Pooideae. In: T.R. Soderstrom, K.W. Hilu, C.S. Campbell, and M.E. Barkworth, eds. Grass systematics and evolution. Smithsonian Institution, Washington, DC. Pp. 265–276.

MACFARLANE, T., N. LANDER, and A. CHAPMAN. 1995. Angiosperm family identification and the flora of Australia, a comment. Austra. Syst. Bot. Soc. Newsl. 83:3–8.

MACFARLANE, T.D. and L. WATSON. 1982. The classification of Poaceae subfamily Pooideae. Taxon 31:178–203.

O'BRIEN, C.W. and I.S. ASKEVOLD. 1992. Systematics and evolution of weevils of the genus *Bagous* Germar (Coleoptera: Curculionidae), I. Species of Australia. *Trans. Amer. Entomol. Soc.* 118:331–452.

O'BRIEN, C.W., I.S. ASKEVOLD, and K. MORIMOTO. 1994. Systematics and evolution of weevils of the genus *Bagous* Germar (Coleoptera: Curculionidae) II. Taxonomic treatment of the species of Japan. *Esakia* 34:1–73.

O'BRIEN, C.W., and I.S. ASKEVOLD. (In preparation). Systematics and evolution of weevils of the genus *Bagous* Germar (Coleoptera: Curculionidae) IV. Taxonomic treatment of the species of the Indian subcontinent. *Oriental Insects?*

PETERSON, P.M., R.D. WEBSTER, and J. VALDES-REYNA. 1997. Genera of New World Eragrostideae (Poaceae: Chloidoideae). Smithsonian Contr. Bot. 87:1–50.

RAMBOLD, G. and G. HAGEDORN. 1998. The distribution of selected diagnostic characters in the Lecanorales. Lichenologist 30:473–487.

ROMI, R., G. PONTUALE, and G. SABATINELLI. 1997. Le zanzare Italiane: Generalitá e identificazione degli stadi preimaginali (Diptera, Culicidae). Fragm. Entomol. Roma 29(Suppl.) 1–141.

SHAW, R.B. and R.D. WEBSTER. 1987. The genus *Eriochloa* (Poaceae: Paniceae) in North and Central America. Sida 12, 165–207.

SIMON, B.K 1992. A revision of the genus *Aristida* (Poaceae) in Australia. Austral. Syst. Bot. 5:129–226.

SMITH, U.R. 1998. Systematic analysis of Oceanopapaver, a "living fossil." Amer. J. Bot. 85:157.

SMITH, U.R. and L.J. HICKEY. 1994. A capsular fruit from the latest Cretaceous of the Western Interior United States. Amer. J. Bot. 81:102–103.

SNYDER, D.E. and R.T. MUTH. 1990. Descriptions and identification of Razorback, Flannelmouth, White, Utah, Bluehead, and Mountain Sucker Larvae and early juveniles. Colorado Div. Wildlife Tech. Publ. No. 38.

TAYLOR, R.W. 1978. A taxonomic guide to the ant genus *Orectognathus* (Hymenoptera: Formicidae). CSIRO Austral. Div. Entomol. Rep. 3:1—11, + microfiche.

TAYLOR, R.W. 1990. New Asian ants of the tribe Basicerotini, with an on-line computer interactive key to the twenty-six known Indo-Australian species (Hymenoptera: Formicidae: Myrmicinae). Invertebr. Taxon. 4:397–425.

TEIXEIRA, A.R. 1994. Genera of Polyporaceae: An objective approach. Bol. Chacara Bot. Itu 1:1–91.

THRANE, U. 1991. FUSKEY, an interactive computer key to common *Fusarium* species. Mycotoxin Res. 7A, Part 1:50–53.

TISSOT-SQUALLI HOUSSAINI, M.L. 1997. Monographische bearbeitung von *Paepalanthus* subgenus *Platycaulon*. J. Cramer, Berlin and Stuttgart.

VAN DEN BORRE, A. and L. WATSON. 1994. The infrageneric classification of *Eragrostis* (Poaceae). Taxon 43:383–421.

VEZEY, E.L., I.K. FERGUSON, and J.J. SKVARLA. 1991. Computer-generated key to pollen of tribe Swartzieae (Leguminosae). Rev. Paleobot. Palynol. 68:289–302.

VELDKAMP, J.F. 1996. Revision of *Panicum* and *Whiteochloa* in Malesia (Gramineae-Paniceae). Blumea 41:181–216.

WALLACE, C.C. and M.J. DALLWITZ. 1981. Key to species of the coral genus *Acropora* from the Great Barrier Reef (Prototype). James Cook University, Townsville.

WATSON, L. 1971. Basic taxonomic data: The need for organization over presentation and accumulation. Taxon 20:131–136.

WATSON, L. 1981. An automated system of generic descriptions for Caesalpinioideae, and its application to classification and key-making. In: R.J. Polhill and P.H. Raven, eds. Advances in legume systematics. Pp. 65–80 + microfiche.

WATSON, L. 1987. Automated descriptions of grass genera. In: T.R. Soderstrom, K.W. Hilu, C.S. Campbell, and M.E. Barkworth, eds. Grass systematics and evolution. Smithsonian Institution, Washington, DC. Pp. 343–355.

WATSON, L., S.G. AIKEN, M.J. DALLWITZ., L.P. LEFKOVITCH, and M. DUBÉ. 1986. Canadian grass genera: Keys and descriptions in English and French from an automated data bank. Canad. J. Bot. 64:53–70 (with 2 microfiches).

Watson, L., H.T. Clifford, and M.J. Dallwitz. 1985. The classification of the Poaceae: Subfamilies and supertribes. Austral. J. Bot. 33:433–484.

Watson, L., and M.J. Dallwitz. 1983. Genera of Leguminosae-Caesalpinioideae: Anatomy, morphology, classification and keys. Research School of Biological Sciences, Australian National University, Canberra.

Watson, L. and M.J. Dallwitz. 1985. Australian grass genera: Anatomy, morphology, keys and classification. 2nd ed. Research School of Biological Sciences, Australian National University, Canberra.

Watson, L., and M.J. Dallwitz. 1988. Grass genera of the world: Illustrations of characters, classification, interactive identification, information retrieval. With microfiches, and floppy disks for MS-DOS microcomputers. Research School of Biological Sciences, Australian National University, Canberra.

Watson, L., and M.J. Dallwitz. 1991. The families of angiosperms: Automated descriptions, with interactive identification and information retrieval. Austral. Syst. Bot. 4:681–695.

Watson, L., and M.J. Dallwitz. 1992. The grass genera of the world. CAB International, Wallingford.

Watson, L., and M.J. Dallwitz. 1992+. Grass genera of the world: Descriptions, illustrationsi, and information retrieval; including synonyms, morphology, anatomy, physiology, cytology, classification, pathogens, world and local distribution, and references. http://biodiversity.uno.edu/delta/

Watson, L., and M.J. Dallwitz. 1992+. The families of flowering plants: Descriptions, illustrations, identification, and information retrieval. http://biodiversity.uno.edu/delta/

Watson, L., and M.J. Dallwitz. 1993. The families of flowering plants. CD-ROM, Version 1.0 for MS-DOS. CSIRO Publications, Melbourne.

Watson, L. and M.J. Dallwitz. 1994. The grass genera of the World. 2nd ed. CAB International, Wallingford.

Watson, L., M.J. Dallwitz, A.J. Gibbs, and R.J. Pankhurst. 1988. Automated taxonomic descriptions. In: D.L. Hawksworth, ed. Prospects in Systematics. Clarendon Press, Oxford. Pp. 292–304.

Watson, L., M.J. Dallwitz, and C.R. Johnston. 1986. Grass genera of the world: 728 detailed descriptions from an automated database. Austral. J. Bot. 34:223–230.

Watson, L., G.E. Gibbs Russell, and M.J. Dallwitz. 1989. Grass genera of southern Africa: Interactive identification and information retrieval from an automated data bank. S. African J. Bot. 55:452–463.

Watson, L. and P. Milne. 1972. A flexible system for automatic generation of special-purpose dichotomous keys, and its application to Australian grass genera. Austral. J. Bot. 20, 331–52.

Webster, R.D. 1984. A revision of the genus *Digitaria* Haller (Paniceae: Poaceae) in Australia. Brunonia 6:131–124.

Webster, R.D. 1987. The Australian Paniceae (Poaceae). (J. Cramer: Berlin and Stuttgart.)

Webster, R.D. 1987. Taxonomy of *Digitaria* section *Digitaria* in North America (Poaceae: Paniceae). Sida 12:209–222.

Webster, R.D. 1988. Genera of the North American Paniceae (Poaceae: Panicoideae). Syst. Bot. 13:576–609.

Webster, R.D. 1992a. Old World genera of the Paniceae (Poaceae: Panicoideae). Sida 15:9–40.

Webster, R.D. 1992b. Character significance and generic similarities in the Paniceae (Poaceae: Panicoideae). Sida 15:185–213.

Webster, R.D. and S.L. Hatch. 1990. Taxonomy of *Digitaria* section Aequiglumae (Poaceae: Paniceae). Sida 14:145–167.

Webster, R.D., J.H. Kirkbride, and J.V. Reyna. 1989. New World genera of the Paniceae (Poaceae: Panicoideae). Sida 13:393–417 + microfiche.

Webster, R.D. and J.V. Reyna. 1988. Genera of Mesoamerican Paniceae (Poaceae: Panicoideae). Sida 13:187–221.

Webster, R.D., and R. Shaw. 1995. Taxonomy of the native North American species of *Saccharum* (Poaceae: Andropogoneae). Sida 16:551–580.

Xu Zhu, L. Watson, M.J. Dallwitz, Shi Wen Gui, Nie Su Mei, Ma Yu Bau, and Zhang Hai-yan. 1997. Grass genera of China: Automated taxonomic descriptions and making keys. In Chinese. Nei Monggol People's Publishing House, Huhhot, China.